I0064513

Nanomaterials in Glucose Sensing

Krishna Burugapalli[1]

Ning Wang[1,2]

Jakub Trzebinski[1,3]

Wenhui Song[2]

Anthony Cass[3]

[1]001, Heinz Wolff Building, Brunel Institute for Bioengineering, Brunel University, Uxbridge, London, UB8 3PH, UK, Tel: +44(0)1895266926, E-mail: krishna.burugapalli@brunel.ac.uk
[2]Wolfson Centre for Materials Processing, Brunel University, Uxbridge, London, UB8 3PH, UK
[3]Department of Chemistry and Institute of Biomedical Engineering, Imperial College London, South Kensington Campus, Exhibition Road, London, SW7 2AZ, UK

© 2014, ASME, 2 Park Avenue, New York, NY 10016, USA (www.asme.org)

All rights reserved. Printed in the United States of America. Except as permitted under the United States Copyright Act of 1976, no part of this publication may be reproduced or distributed in any form or by any means, or stored in a database or retrieval system, without the prior written permission of the publisher.

Co-published by Momentum Press, LLC, 222 E. 46th Street, #203, New York, NY 10017, USA (www.momentumpress.net)

INFORMATION CONTAINED IN THIS WORK HAS BEEN OBTAINED BY THE AMERICAN SOCIETY OF MECHANICAL ENGINEERS FROM SOURCES BELIEVED TO BE RELIABLE. HOWEVER, NEITHER ASME NOR ITS AUTHORS OR EDITORS GUARANTEE THE ACCURACY OR COMPLETENESS OF ANY INFORMATION PUBLISHED IN THIS WORK. NEITHER ASME NOR ITS AUTHORS AND EDITORS SHALL BE RESPONSIBLE FOR ANY ERRORS, OMISSIONS, OR DAMAGES ARISING OUT OF THE USE OF THIS INFORMATION. THE WORK IS PUBLISHED WITH THE UNDERSTANDING THAT ASME AND ITS AUTHORS AND EDITORS ARE SUPPLYING INFORMATION BUT ARE NOT ATTEMPTING TO RENDER ENGINEERING OR OTHER PRO-FESSIONAL SERVICES. IF SUCH ENGINEERING OR PROFESSIONAL SERVICES ARE REQUIRED, THE ASSISTANCE OF AN APPROPRIATE PROFESSIONAL SHOULD BE SOUGHT.

ASME shall not be responsible for statements or opinions advanced in papers or . . . printed in its publications (B7.1.3). Statement from the Bylaws.

For authorization to photocopy material for internal or personal use under those circumstances not falling within the fair use provisions of the Copyright Act, contact the Copyright Clearance Center (CCC), 222 Rosewood Drive, Danvers, MA 01923, tel: 978-750-8400, www.copyright.com.

Requests for special permission or bulk reproduction should be addressed to the ASME Publishing Department, or submitted online at: http://www.asme.org/kb/books/book-proposal-guidelines/permissions

ASME Press books are available at special quantity discounts to use as premiums or for use in corporate training programs. For more information, contact Special Sales at infocentral@asme.org

A catalog record is available from the Library of Congress.

Print ISBN: 978-0-7918-6027-4
ASME Order No. 860274
Electronic ISBN: 978-1-60650-642-4

Series Editors' Preface

Biomedical and Nanomedical Technologies (B&NT)
This **concise** monograph series focuses on the implementation of various engineering principles in the conception, design, development, analysis and operation of biomedical, biotechnological and nanotechnology systems and applications. The primary objective of the series is to compile the latest research topics in biomedical and nanomedical technologies, specifically devices and materials.

Each volume comprises a collection of invited manuscripts, written in an accessible manner and of a concise and manageable length. These timely collections will provide an invaluable resource for initial enquiries about technologies, encapsulating the latest developments and applications with reference sources for further detailed information. The content and format have been specifically designed to stimulate further advances and applications of these technologies by reaching out to the non-specialist across a broad audience.

Contributions to *Biomedical and Nanomedical Technologies* will inspire interest in further research and development using these technologies and encourage other potential applications. This will foster the advancement of biomedical and nanomedical applications, ultimately improving healthcare delivery.

Editor:
Ahmed Al-Jumaily, PhD, Professor of Biomechanical Engineering & Director of the Institute of Biomedical Technologies, Auckland University of Technology.

Associate Editors:
Christopher H.M. Jenkins, Ph, PE, Professor and Head, Mechanical & Industrial Engineering Department, Montana State University.

Guy M. Genin, PhD, Associate Professor of Mechanical Engineering and Materials Science, Washington University in St. Louis, and Associate Professor of Neurological Surgery, Washington University School of Medicine.

Feng Xu, PhD, Professor, The Key Laboratory of Biomedical Information Engineering of Ministry of Education, School of Life Science and Technology, and Director, XJTU Biomedical Engineering & Biomechanics Center, Xi'an Jiaotong University, China.

Contents

Contents

Abstract

The smartness of nano-materials is attributed to their nanoscale and subsequently unique physicochemical properties and their use in glucose sensing has been aimed at improving performance, reducing cost and miniaturizing the sensor and its associated instrumentation. So far, portable (handheld) glucose analysers were introduced for hospital wards, emergency rooms and physicians' offices; single-use strip systems achieved nanolitre sampling for painless and accurate home glucose monitoring; advanced continuous monitoring devices having 2 to 7 days operating life are in clinical and home use; and continued research efforts are being made to develop and introduce increasingly advanced glucose monitoring systems for health as well as for food, biotechnology, cell and tissue culture industries. Nanomaterials have touched every aspect of biosensor design and this monograph reviews their role in the development of advanced technologies for glucose sensing, and especially for diabetes management.

Research shows that overall, nanomaterials help address the problems with conventional optical and electrochemical biosensors, by enhancing the preferential detection of glucose or its oxidation products through better electron transfer kinetics, sensitivity and response time, while lowering the operating over-voltages for energy efficiency and avoid interference. The reproducible production of nano-materials and nano-structures at low cost is vital for the successful development of nano-technologies for glucose sensing. Several products, especially, home glucose monitoring devices, use nano-materials, but the need for reliable long-term CGM is still unmet. Nano-materials and nano-technologies have an important role in achieving the long-awaited CGM technology.

1. Introduction

Glucose is the fundamental fuel for cells of most living organisms. It is produced by plants and some prokaryotes through photosynthesis, wherein solar energy is absorbed, used to generate reducing equivalents ('light reactions') and then fix carbon dioxide as glucose ('dark reactions'). Excess glucose is stored as polymeric forms, starch in plants and glycogen in animals. To utilize glucose as the source of energy, living organisms (from bacteria to humans) evolved glucose specific metabolic enzymes. Through processes, namely, aerobic respiration, anaerobic respiration or fermentation, cells use glucose to generate energy for production of other sugars, proteins, lipids, vitamins and nucleic acids that orchestrate the genesis, structure and function of all living systems. Proteins and lipids are also used by organisms not capable of photosynthesis to produce glucose through a process called gluconeogenesis. Under anaerobic and dark conditions, deep-sea bacteria developed chemosynthesis as alternative to photosynthesis. They use molecules such as carbon dioxide or methane as carbon source and inorganic molecules such as hydrogen gas, or hydrogen sulfide as a source of reducing equivalents to produce glucose. Commercially, glucose is manufactured through enzymatic hydrolysis of starch derived from crops.

Essential for most life processes, glucose is one of the most abundant molecules on earth. Almost entirely generated and utilized through natural processes, human intervention through large-scale farming (crops), domestication (animals), science and technology have only intensified its customised production to sustain an expanding human population. Thus, glucose has had indirect socio-economic impacts on all human civilizations. Over millennia, it has been a basic component in food and health, and more recently in biotechnology, sustainable energy and, cell and tissue culture industries. Commercially, glucose is also used as a precursor for the production of molecules such as vitamin C, citric acid, gluconic acid, polylactic acid, sorbitol and ethanol. Such large-scale utilization of glucose necessitates the need for efficient feedback control systems. The best examples of such tightly regulated systems were developed by nature in living organisms, the failure of which, in disease, can be fatal (without human intervention). Further, in the last three decades there has been an exponential increase in the use of glucose in commercial production sectors, requiring tight quality control standards. Thus, glucose sensing has a vital role in effective feedback controlled management of glucose utilizing systems. This monograph reviews the current state of glucose sensing technologies, the need for nanomaterials and their role in pushing the technological boundaries to meet the un-met and the most demanding of glucose sensing needs.

2. Need for glucose sensing

The need for glucose sensing has primarily been driven by the medical condition – diabetes, wherein, the normal glucose metabolism is disturbed. The usual blood glucose level in healthy humans varies between 70 to 120 mg/dL or 4 to 8 mM/L. When the body's feedback control system is impaired in people with diabetes there is a much wider range, 30 to 500 mg/dL or 2 to 30 mM/L glucose. In these cases, insulin, the hormone that promotes glucose uptake by cells, is either not produced in sufficient quantity (Type I or insulin dependent diabetes) or the glucose absorbing cells develop an insensitivity to insulin (Type II or non-insulin dependent diabetes). The opposite effect, wherein hyperinsulinism or lack of counter-regulatory hormones (e.g. Glucagon), abnormally lowers blood glucose levels (hypoglycemia). Persistent hyperglycaemia causes dehydration, long-term cardiovascular complications, damage to the eyes and kidneys, and impaired would healing; while the hypoglycemia often results in fainting, coma, or death [1, 2]. The conditions hyper- or hypo-glycaemia can also occur due to non-diabetic and transient causes such as, pregnancy, medication, stress/trauma, haemorrhage, burns, infections, stroke or alcohol consumption [2, 3]. In either case, it is essential to identify and control the blood glucose levels (glycemic control), to avoid painful and debilitating complications. Tight glycemic control is only possible when reliable, accurate and timely glucose sensing is achieved [1–4].

In addition to the diagnosis of diabetes, glucose sensing has become essential in the process optimization of industrial and research scale microbial and mammalian cell cultures. *In vitro* cell culture processes are complex and difficult to optimize unless a feedback control based on tight monitoring of a wide range of parameters including glucose, lactate, pH and temperature are achieved. An industrial scale example of glucose sensing is represented in microbial fermentation processes. Traditionally, intermittent sampling of fermentation medium is performed, sample processed and used to monitor glucose using a commercial glucose analyser (for example that from Yellow Spring Instruments, Yellow Spring, Ohio, USA) [5]. At a research level, the focus is on developing mammalian cell culture bioreactors integrated with various sensors including glucose biosensors for regenerative medicine and toxicity testing, especially to reduce animal use in these areas. Recent advances in glucose sensing, especially real-time (continuous) glucose monitoring technologies developed for diabetes management, are being explored and utilized for cell culture bioreactors and microfluidic devices [6–9].

3. Signal generation and detection of glucose

Electrochemical and optical methods are commonly utilized to generate and detect measurable signals that are glucose concentration-dependent. Oliver et al. and Pickup et al. describe the principles, advantages and disadvantages of the different optical methods [4, 10], while that of electrochemical methods are reviewed in detail by Heller and Feldman [1].

3.1 Electrochemical methods

Figure 3-1 depicts the major electrochemical methods used to convert the free energy of glucose into an electrical signal. Broadly they can be classified as either non-enzymatic or enzymatic methods. The former are usually non-specific and suffer from signal inaccuracies due to interferents and electrode poisoning, while the latter are more specific and have become the mainstay of clinical and home glucose monitoring.

3.1.1 Direct (non-enzymatic) oxidation or reduction of glucose at an electrode

Direct electrooxidation or reduction of glucose in an electrolytic cell usually require extreme conditions, such as pH > 11, pH < 1 and high voltages, or suffer from electrode poisoning, inhibition and interference at physiological pH (Table 3-1) [1, 11]. The different isomeric (α-, β-) forms of glucose are electrooxidized at a metal electrode, whether they involve intermediates or not, to gluconic acid as the final stable product of a two-electron oxidation of glucose [11–13]. The resulting gluconolactone is reported to have a half-life of 10 min and rate constant of 10^{-3} s^{-1} at pH 7.5 for hydrolysis [11]. Furthermore, the electrooxidation of glucose was shown to have three potential ranges vs. the reversible hydrogen electrode (RHE) of relevance to glucose sensing (Table 3-1) [11, 14]. The first, between 0.15 and 0.35 V, called 'hydrogen region', provides strong oxidation peak potentials unique to glucose, observed primarily with Pt electrode. However, flat platinum electrodes rapidly lose sensitivity to glucose due to poisoning by chloride ions and various organic agents including amino acids, acetaminophen, creatinine, epinephrine, urea and uric acid in physiological solution, and this normally blocks the catalytic sites on platinum, inhibiting glucose oxidation. The second, between 0.4 and 0.8 V vs. RHE, called 'double layer region' lowers the adsorptive capacity of the poisoning agents, but other electroactive (interfering) agents are also oxidized at the electrode. Finally, the third, 'Pt oxide region', between 1.1 and 1.5 V produces a Pt oxide layer, with which glucose reacts, while poisoning products similar to lactones are decomposed by further oxidation. Overall, the lack of selective catalytic activity to glucose and electrode poisoning makes such systems unsuited for direct glucose sensing in complex biological fluids.

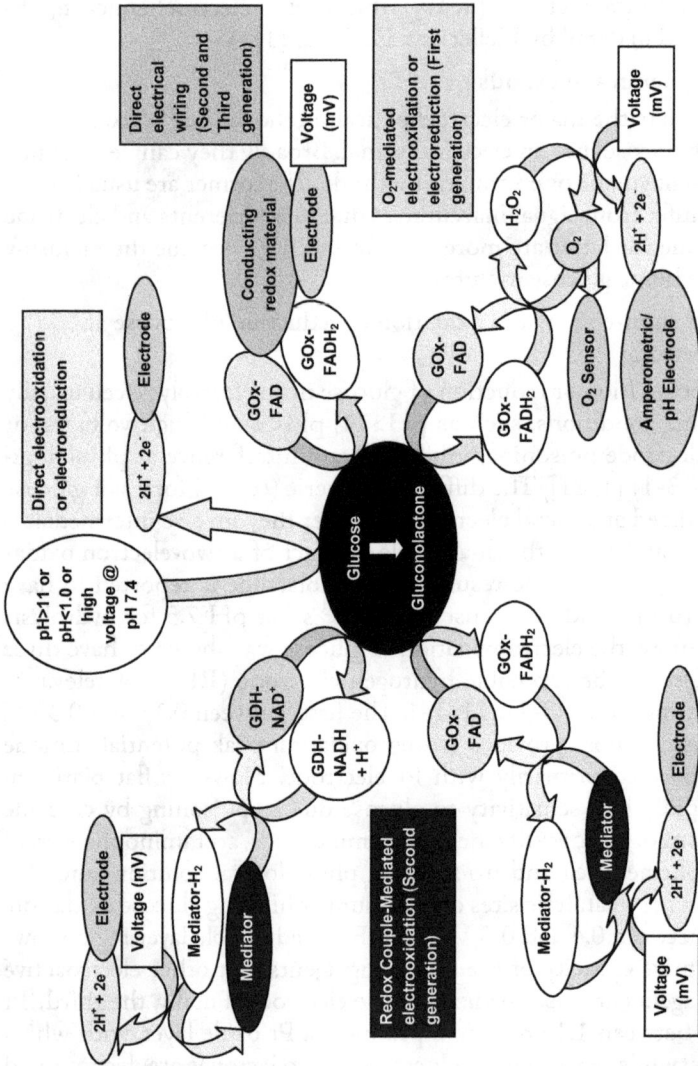

Figure 3-1 Electrochemical methods for glucose detection.

Table 3-1   Direct electro-oxidation of glucose at electrodes [1, 11, 14]

Electrode	Conditions	Disadvantages
Conventional electrodes (e.g. Pt and Au)	pH>11, pH<1, High voltages	Electrode poisoning, Inhibition Interference Extreme conditions
Reversible Hydrogen Electrode (RHE)	0.15 to 0.35 V (hydrogen region)	Electrode poisoning, agents such as chloride ions, acetaminophen, creatinine, epinephrine, urea and uric acid, bind and block catalytic sites on electrodes Selectivity not exclusive to glucose
Reversible Hydrogen Electrode (RHE)	0.4 to 0.8 V (double layer region)	Electrode poisoning is reduced but there is interference due to oxidation of other electroactive agents including ascorbic acid, acetaminophen and uric acid Selectivity not exclusive to glucose
Reversible Hydrogen Electrode (RHE)	1.1 to 1.5 V (Pt oxide region)	High voltages decompose interferents but not glucose, but selectivity not exclusive to glucose

3.1.2   Indirect oxidation or reduction of glucose at an electrode

Indirect methods are based on reactions of glucose with affinity molecules, predominantly enzymes immobilized on electrode surface. Two families of enzymes, namely, glucose oxidases (GOx) and glucose dehydrogenases (GDH), are the most widely studied and used enzymes for glucose sensing [1]. GOx is a dimeric protein with one flavin adenine dinucleotide (FAD) cofactor per subunit strongly attached and buried deep in a protective glycoprotein shell as its redox centre [15, 16]. It is very specific to β-D-glucose and oxidizes about 5000 molecules per second in an electrochemical half-reaction. On the other hand, GDH has pyrroloquinoline quinine (PQQ) as a moderately tightly bound cofactor and oxidizes glucose (as well as other sugars) at about 11,800 molecules per second in an electrochemical half reaction. In the process of oxidising glucose, the cofactors of GOx and GDH get reduced ($FADH_2$ and $PQQH_2$) and require rapid regeneration (oxidation) to sustain turnover. GOx-$FADH_2$ is naturally oxidised by oxygen (O_2), while the reoxidation of GDH-$PQQH_2$ is O_2 independent (Figure 3-1). Two other GDHs – nicotinamide adenine dinucleotide (NAD) and FAD dependent – have also been investigated as they combine the oxygen independence of GDH-PQQ and specificity to glucose of GOx-FAD.

3.1.2.1   First-generation sensors

Glucose sensors are often categorised into three generations based on the enzyme electrochemical coupling used to make the glucose measurement

[17, 18]. The first-generation sensors use the natural O_2 co-substrate to oxidise GOx-$FADH_2$, to generate hydrogen peroxide (H_2O_2) (O_2 – mediated electrooxidation or electroreduction, Figure 3-1). The advantage of this method is that H_2O_2 can be directly oxidized at an anodic working (conventional solid metal) electrode at potentials between +0.6 to +0.7 V vs. saturated calomel (SCE) or silver/silver chloride (Ag/AgCl) reference electrode, which accurately correlates to the oxidised glucose concentration. However, other naturally occurring molecules, such as ascorbic acid, uric acid and acetaminophen, are also oxidized at the electrode between +0.6 to +0.7 V. Another problem for first-generation sensors is the low tissue oxygen concentration (0.1 to 0.3 mM) compared to 2–30 mM of glucose in diabetics [18]. To address these problems, technologies based on nano-structured membranes that prevent interfering agents (details in Section 5) and those that limit the flux of glucose reaching the enzyme (details in Section 6) have been developed.

In addition to H_2O_2 production, changes in oxygen and pH associated with the first generation sensing reaction are measured and related to glucose concentrations. The first commercial clinical analyzer, from Beckman, introduced in 1968, was based on electroreduction of O_2 at a gold cathode, which later was replaced by a polarographic O_2-electrode [1, 19]. Recently, Gough et al. reported the use of GOx and catalase based dual enzyme O_2 sensing system for continuous glucose monitoring (CGM) in a subcutaneous animal model for over a year [20]. Significant research also focused on using field-effect transistor (FET) sensors to monitor pH changes due to the electrooxidation of H_2O_2 and/or the hydrolysis of gluconolactone to determine changes in glucose concentrations. However, pH (potentiometric) sensors for glucose monitoring suffer from inaccuracies due to local tissue parameters effecting pH including the reaction kinetics generating H^+ ions for detection [1].

3.1.2.2 *Second-generation sensors*
Second-generation sensors, typically, have synthetic chemical (artificial) mediators replacing O_2 as co-substrates for oxidising GOx-$FADH_2$ or GDH-$PQQH_2$ (mediator coupled-electrooxidation, Figure 3-1). The mediators have nano-dimensions capable of directly interacting with $FADH_2$ of GOx, and mediate the transfer of electrons to the electrode (details in Section 4.4). They obviate the need for O_2 but suffer from competition from O_2. To circumvent O_2 competition, a large excess of mediator is incorporated, and in many currently available single-use home glucose test strips, GOx is replaced by GDH-PQQ. However, sensors with GDH-PQQ suffer from inaccuracy due to reaction with non-glucose sugars, such as maltose, xylose, and galactose (found in certain drug and biologic formulations, or can result from the metabolism of a drug or therapeutic product), often leading to falsely elevated blood glucose levels prompting deleterious therapies [21]. In

addition to O_2 independence, second-generation sensors, depending on the structure of the mediator, require lower electrode potentials for measurement compared to that for direct oxidation of H_2O_2, thus avoiding the common interferents from blood/tissue oxidising at the electrode. In addition, a restrictive membrane to slow down the diffusion of glucose is often not required. However, a membrane to entrap or immobilize the mediator and enzyme at the electrode surface is often essential for using the second-generation sensors in CGM [18]. Among the immobilizing membranes, Heller et al. demonstrated the efficacy of redox polymers leading to the design of commercial second-generation sensors [22–24].

For the successful design of a commercial second-generation sensor for CGM, it is essential to avoid the free diffusion of artificial mediators into the surrounding medium. To this goal, Heller's group engineered novel redox hydrogels that immobilize mediator redox centres through covalent attachment to insoluble, but water swollen, crosslinked hydrophilic polymers. The redox hydrogels are used to envelop GOx enzyme, such that the enzyme's reaction centre is electrically connected directly to the electrode surface [1, 22, 23, 25]. Recently, Willner's group also developed a similar system wherein GOx is covalently bound to Au NPs that are crosslinked into multiple layers electrically wiring the redox centres of GOx directly on to Au electrode surface [26, 27]. An added advantage of such direct electrical wiring systems is that the crosslinked redox networks connect multiple enzyme layers, thus increasing the glucose measuring current by 10 to 100 fold and current densities of glucose electrooxidation on conventional solid electrodes can exceed 1 mA/cm² at potentials between 0.0–0.1 V *vs.* Ag/AgCl [1].

3.1.2.3 Third-generation sensors

The futuristic third-generation sensors aim to avoid the redox mediators, and achieve direct transfer of electrons from glucose to electrode via the redox centre of the enzyme. The resulting glucose sensors, said to be reagentless, and are expected to have a low operating potential that is close to the redox potential of the enzyme [17]. The $FADH_2$ redox centre of the conventional GOx enzyme is know from the crystal structure to be buried at a depth of 13–15 Å, due to which, direct electron transfer from GOx to conventional solid state electrodes is reported to be much too slow [1, 16]. Hence, the design of third generation glucose sensors is expected to either utilize other glucose oxidising enzymes or explore new electrode materials. Wang, and Chaubey and Malhotra have provided brief reviews of the different electrode materials being investigated for the design of third-generation sensors [17, 28]. An example for one such electrode materials is tetrathiafulvalene-tetracyanoquinodimethane (TTF-TCNQ), which has been widely reported for the development of conducting organic salt electrodes for direct transfer of electrons from glucose to the electrode through the redox centre of GOx [17].

3.2 Optical methods

Spectroscopic measurements of direct or indirect interactions of different properties of light with glucose molecules alone; colour generating glucose-dye complexes; or fluorescence systems based on glucose–receptor molecule complexes form the firm basis for the optical methods.

3.2.1 Direct spectroscopic methods

A direct method to measure blood glucose levels is to irradiate tissues such as skin, eye, or whole blood, with different forms of light, and detect the resulting absorption, transmission, reflection, scattering or emission signals [4]. Mid infrared (MIR) radiation with wavelengths (λ) between 2.5 to 50 μm and near infrared (NIR, 0.7 to 1.4 μm) are widely studied for this purpose. Although the tissue penetration depth for MIR is much low (superficial compared to NIR that reaches up to the subcutaneous space), MIR was successfully used in glucose measurements, ex vivo, in whole blood in intensive care units [29, 30]. Other commonly investigated forms of light for glucose detection include, plane polarized, coherent (light in which emitted photons are synchronised in time and space), single wavelength (Raman) and He-Ne laser light [31–34]. Despite being non-invasive, these methods often fail to reach the standards for continuous glucose monitoring (CGM) in vivo, because of their, usually, >20% error in glucose measurements [3, 35]. The large error is caused by factors including interferents; heterogeneity of tissue; temperature; heterogeneous distribution of glucose between cells, interstitium and blood; changes in water, free fatty acids, chylomicrons, fats or protein content in tissue; motion artefacts (including blood flow); and drug treatments [3, 4, 10]. Often the glucose measuring signal is weak (esp. for Raman spectroscopy), and to quantify the single solute, the signal requires screening of multiple wavelengths and multivariate statistical analysis with calibration [4].

A promising method for direct spectroscopic measurement of glucose is the combination of selective partitioning and surface-enhanced Raman spectroscopy (SERS). SERS provides a 10^6 to 10^{14} times enhancement of the Raman spectral signal, when a Raman-active molecule comes within the surface plasmonic field (0 to 2 nm) on roughened metal surfaces. This not only allows ultrasensitive detection but also molecular structure identification. van Duyne's group have extensively researched the application of SERS for in vitro and in vivo assays of glucose [36]. They functionalised Ag and Au surfaces with tri(ethylene glycol)-terminated alkanethiol, which layer reversibly partitions glucose within the surface plasmonic field, <2.8 nm from the surface, and its finger print Raman spectrum allows selective and ultrasensitive SERS detection, even in the presence of other sugars and lactate.

3.2.2 Indirect spectroscopic methods

Glucose can also be measured by indirect methods, where glucose molecules modify a specific tissue property that can be detected. Photo-acoustic spectroscopy are two such methods used for measuring the glucose-mediated decrease in specific heat capacity and the glucose-red blood cell interactions causing concentration-dependent changes in dielectric properties of tissue respectively [37, 38]. These methods also suffer from similar problems of lack of specificity faced by the above-described direct methods.

3.2.3   Glucose-specific affinity-based methods

By far the most efficient, reliable and practical way to measure glucose is to use glucose specific interactions [1, 4]. Most fluorescence based optical methods utilize the affinity sensing principle, where glucose competitively displaces a fluorescently labelled binding agent (e.g., dextran, α-methyl mannoside or glycated protein) from a receptor, specific to both glucose and binding agent, resulting in a concentration-dependent change in fluorescence emission or quenching signal [4, 10]. The glucose-specific affinity molecules can broadly be classified into four categories: Concanavalin A (Con A), enzyme based (glucose oxidase/dehydrogenase, hexokinase), synthetic boronic acid derivatives, and bacterial glucose specific proteins. While the former two are well established, newer and better molecules of the latter two categories are being developed for glucose sensing. Further, an advanced and extremely sensitive fluorescent method is the fluorescence resonance energy transfer (FRET) based on Angstrom level dipole-dipole interactions between fluorescence donor and acceptor molecules leading to a decrease in fluorescence intensity or lifetime of a donor molecule. In spite of their huge promise, fluorescence based glucose detection methods are yet to reach the clinics for CGM *in vivo*.

4. Nanomaterials generating or enhancing glucose measuring signals

Nano-materials, by virtue of their nanoscale dimensions, make it possible to extend glucose sensing technologies beyond the limitations of conventional electrodes. In other words, the advent of nanotechnology opened up huge scope for development of miniature and CGM systems having wider detection limits and better sensitivity. Typically, nanomaterials have high surface area to volume ratios and high surface activity, due to which and depending on the ambient environment, often the same material behaves as electrical or thermal conductor, semi-conductor, optical or mechanical transducer. Hence, their unique catalytic and signal transducing abilities find them a variety of sensing applications.

4.1 Non-enzymatic electrochemical glucose sensing

The mechanisms of direct electrooxidation of glucose at conventional electrodes are summarised in Section 3.1.1 and Table 3-1. Glucose, a monosaccharide, can indeed be directly electrooxidized at conventional electrodes. However, its rate of oxidation is much slower than common interfering substances such as uric acid and ascorbic acid in biological fluids. Though the normal concentration of glucose in blood (3 to 8 mM) is much higher than the interfering substances (0.1 mM), the faradic current generated by oxidation of interfering agents is much higher than that for glucose [11]. Attempts have been made to avoid interference by lowering the overpotential for oxidation of glucose through the development of electrodes based on alloys of Pt, Pb, Au, Pd, Rh, Cu, Mn, Ni, Fe, Ag etc. Glucose is oxidized at Pt_2Pb alloy surface at remarkably low potentials (compared to pure platinum) at which the electrode is insensitive to common interfering substances [39]. Nevertheless, they failed to prevent electrode poisoning by chloride ions and other electroactive components abundant in physiological solutions as well as by the various glucose oxidation products. The advent of nanotechnology provided new hope for non-enzymatic glucose sensing, and combining nanomaterials with conventional electrodes not only avoided interference from interfering substances, but also decreased or prevented electrode poisoning.

Electrodes modified with metallic nano-particles (NPs) provide enhanced electroactive surface area, mass transport and catalysis [40–44]. Introduction of Ag on Au electrodes in the form of Ag-Au alloy NPs lowered the glucose oxidation potential to 0.3 V [40]. Casella et al. demonstrated that dispersion of Cu micro-particles on Au electrode yielded a stable multifunctional amperometric sensor that can detect glucose using constant-potential and pulsed-potential amperometric detection methods, both under static and forced flow hydrodynamic conditions, with a detection limit of 0.8 pM of glucose at 0.35 V [41]. Recently, Ag_2O nanowalls consisting of densely packed nanoplates on a Cu substrate were shown to have ultrahigh sensitivity to glucose (298.2 μA/mM) at an applied potential of 0.4 V. The Ag_2O-Cu nanowalls modified electrode was not only successful at eliminating interferences from uric acid and ascorbic acid, but also fructose during the catalytic oxidation of glucose [42]. Higher sensitivities to glucose were also obtained using CuO based materials, e.g., porous CuO films on Pt/Ti/Si substrate showed a sensitivity of 2.9 mA/mM.cm² at 0.65 V [43]. Towards prevention of electrode fouling, Xu et al. successfully utilized dimethylglyoxime functionalized Cu-NPs [44]. Further, the incorporation of NPs on electrode surface by itself prevented electrode poisoning, usually, by protecting the electrode's catalytic sites [11, 45]. However, most of these systems require alkaline conditions for sensing glucose.

Carbon-based materials such as carbon paste, diamond-like carbon, graphite, graphene and carbon nanotubes (CNTs) are widely being explored for glucose sensing, often in combination with NPs. Among the carbon materials, CNTs are widely researched due to their exceptional chemical and physical properties including high surface to volume ratios, chemical stability, electrical conductivity and electron transfer kinetics. Different combinations of carbon based materials and/or NPs, with or without polymer fillers are being tested as coatings on common electrodes such as Pt, Au and glassy carbon for non-enzymatic glucose sensing. Thin films of Ni-NPs with graphite-like carbon were used to detect various sugars including glucose at 0.4 V vs. Ag/AgCl having a detection limit for glucose of 20 nM [46]. Ye et al. first demonstrated the direct oxidation of glucose at multi-wall CNTs (MWNTs) at 0.2 V giving a reproducible sensitivity of 4.36 μA/mM.cm² [47]. Later, freestanding single-wall CNTs (SWNTs) and re-structured CNT forests were also successfully used as stable, highly sensitive and fast responding non-enzymatic glucose sensors [48, 49]. Cu NPs combined with CNTs showed sensitivities as high as 2.19 mA/mM.cm² at about 0.65 V [50–52]. NPs of Au, Cu, Pt, Pd, Pt-Pb alloy, BO and MnO_2 were also tested in combination with CNTs [53–59]. However, the electrocatalytic effects of almost all of the above described nanomaterial coatings are observed in basic conditions usually at pH>9. More recently, due to the large surface area of graphene and its oxides, as well as due to the fast electron transfer between graphene and analyte molecules, several nonenzymatic glucose biosensors based on graphene-modified GCE have been developed in combination with metal/metal alloy or inorganic nanoparticles, such as Ni(II)-quercetin [60], PtNi alloy NPs [61], NiO [62], Ni(OH)s [63] and graphene-Co_3O_4 composites [64]. Among all the graphene-based electrodes, the graphene-Co_3O_4 composite sensor showed the lowest detection limit of 25 nM with a high sensitivity of 3.39 mA mM⁻¹ cm⁻².

One of the simplest, yet effective, enzyme-free sensor is based on a nanoporous Pt electrode, whose surface nanopores have diameters in the range 2–50 nm [65, 66]. This electrode is reported to selectively enhance the faradic current of the normally sluggish glucose oxidation reaction through a kineti-cally controlled electrochemical mechanism. Interfering agents – ascorbic acid and acetaminophen gave negligible response, while signal amplification was observed for glucose oxidation at the nanoporous Pt electrode in PBS, pH 7.4 and 37°C in the presence of 0.15 M NaCl [65]. Thus nanoporosity of Pt not only reversed the analyte selectivity, but also prevented electrode poisoning by chloride ions. Unlike other non-enzymatic glucose biosensors, the elimination of interference from common interfering agents at the nano-porous Pt electrode is not a function of applied potential but of the dynam-ics of mass transport at the nanoporous electrode surface [11]. Compared to the enzymatic sensors, the nanoporous Pt sensor is mechanically and chemically stable, resistant to humidity and temperature, has longer storage shelf-time, allows thermal and chemical sterilization process, and is favour-able for mass production due to simple quality control [11].

Irrespective of the advances in non-enzymatic electrochemical glucose sensing technologies, exclusive selectivity to glucose in complex biological fluids is still very difficult to achieve. Usually, these sensors suffer competi-tion from non-glucose sugars. For example, a potentiometric sensor based on a non-enzymatic affinity molecule, boronic acid, showed higher sensitiv-ity to fructose than glucose [67]. Due to the requirement for alkaline con-ditions, most of the nanomaterial based systems described above are not suitable for implantable CGM systems. Instead they can find applications ex vivo, especially that following sophisticated sample processing by micro-fluidic systems [11, 68]. To conclude, nanomaterials did provide innovative non-enzymatic electrochemical glucose sensors, and are yet to reach the clinics.

4.2 Non-enzymatic optical glucose sensing

Traditional optical systems utilize a dye interacting with glucose or its products in affinity reactions resulting in glucose concentration dependant change in colour. Typically, the dye is irreversibly changed in the reactions. Such systems have been the laboratory gold standards for glucose colorimet-ric assays, and their influence on glucose sensing research was reflected in the first single use home glucose test strips [18]. But to be useful in CGM, the sensing reactions need to be reversible and selective. Further, the opti-cal reading systems need to be in correct spatial orientation and optimal proximity to the colour changes. The instrumentation is usually bulkier and more expensive than electrochemical systems. Many such technical chal-lenges are being addressed through the rapid advances in nanotechnology and instrumentation, which, in future, could lead to the availability of com-mercial optical CGM systems [18].

Reversible affinity interactions of glucose with molecules such as Con A, boronic acid derivatives, and bacterial glucose/galactose binding proteins, have been the main-stay for non-enzymatic optical glucose sensing systems.

The affinity molecules themselves are of nano-size and their molecular scale binding events with glucose are used to trigger an optical signal. To prevent leaching and toxicity, the affinity molecules should either be contained by a membrane or immobilized within an insoluble polymeric network. Ballerstadt et al, used a cellulose based microfiltration membrane with molecular mass cut off 6–8 kDa to prevent the leakage of the toxic Con A-dye conjugate [69]. The membrane not only prevents the leaching of the affinity molecules, but also helps widen the linear glucose detection rate by limiting its diffusion. Similar approach was utilized by Stein et al., to limit the diffusion of glucose while entrapping and protecting optically active microspheres for glucose detection [70, 71]. Using layer-by-layer (LBL) deposition Stein et al. assembled nanometre size (12 nm) layers of poly(allylamine) hydrochloride (PAH) and poly(sodium-4-styrenesulfonate) (PSS) on glucose sensitive microspheres and by varying the assembly conditions and the number of layers adjusted the sensitivity to cover the hypo-, normo- and hyperglycemic levels.

Among various types of optical glucose biosensors, 'smart tattoos' seem to be promising for minimally invasive glucose monitoring technology [71–75]. Light emission from these fluorescent microspheres varies from molecules in different states making them very sensitive and with rapid (approximately 120 s) response to glucose concentration. Various techniques/dyes were used to fabricate glucose sensitive microspheres including Pt(II) octaethylporphine (PtOEP) [70], fluorescein isothiocyanate (FITC)-dextran (FD) and tetramethyl-rhodamine isothiocyanate (TRITC)-Con A [76] or other glucose binding proteins (GBP) [77]. It is important to notice that, in order for these sensors to function properly, a membrane has to be employed to limit the diffusion of glucose and/or entrap and protect the microspheres.

Furthermore, the glucose affinity molecules along with flurophores were also incorporated into hydrogels for optical glucose sensing. For example, Alexeev et al. used photonic properties of boronic acid based hydrogels to quantify glucose concentration [78]. Similarly, the red shifts of phenylboronic acid containing hydrogels for holographic glucose sensors were extensively studied by Lowe's group [79].

The better efficacy and size dependant fluorescence emission of quantum dots (semiconductor nanocrystals) than organic fluorophores has also been exploited for glucose sensing [80]. Cordes et al. used fluorescent CdSe/ZnS quantum dots in combination with boronic acid modified viologens to sense glucose in aqueous solution, having a linear detection range of 2.5 to 20 mM [81]. Con A-conjugated QDs together with thiolated beta-cyclodextrin-modified AuNPs were developed by Tang et al. for use as hyperefficient FRET nanobiosensors for glucose sensing. These sensors had a linear detection range of 0.1 to 50 µM, a detection limit of 50 nM and an excellent selectivity to glucose compared to other sugars [82].

4.3 Indirect enzymatic glucose sensing

Enzymatic glucose sensing systems can broadly be categorised into direct or indirect based on whether the electrons from glucose oxidation by enzymes are catalytically transferred directly to an electrode or indirectly through further redox reactions involving the enzyme's electroactive reactants or products. The indirect means are typically based on O_2 mediated oxidation of glucose by GOx (O_2 mediated electrooxidation or electroreduction, Figure 3-1). Measuring O_2 depletion, H_2O_2 production or pH changes associated with enzymatic reactions (usually first generation sensors) are the commonly utilized indirect methods. In addition, optical methods for indirect measurement of glucose using nanomaterials and enzymatic reactions have been developed, which are covered separately at the end of this section.

4.3.1 Measuring O_2 depletion

Typically, in a GOx catalyzed oxidation of glucose, depending on the presence or absence of catalase enzyme, half or one O_2 molecule respectively, is consumed, leading to a stoichiometric depletion in O_2 concentration in relation to glucose. The changes in the concentration of O_2 are measured using Pt, indium tin oxide, gold or carbon paste electrodes [20, 83–90]. The simplest and most common amperometric O_2 sensor uses a Clark oxygen electrode constituting a platinum cathode where oxygen is electroreduced vs. Ag/AgCl reference electrode at −0.6 V. However, due to the low concentrations of dissolved O_2 (<0.2 mM) in relation to glucose (>2 mM) in the body, it is essential to incorporate smart membranes on electrodes to accurately measure the above said O_2 depletion. Further these systems usually employ a dual electrode system, one with and the other without GOx, and the difference in O_2 concentrations measured between the two electrodes is related to glucose concentration. Suzuki et al. developed miniature micromachined sensor arrays that measure multiple analytes. The miniature sensors utilized hydrophobic O_2 permeable membranes based on a photoresist (OMR-83) or silicone combined with hydrophilic agarose gel or SiO_2 to achieve 90% response currents within one minute and a glucose detection range of 0.02 to 1.4 mM [84–86]. Mass transport-limiting films, such as polyurethane or polycarbonate, were also used to tailor the oxygen/glucose permeability ratio [90–92]. Gough et al. utilized a two dimensional cylindrical electrode system with a hydrophobic polydimethylsiloxane (PDMS) membrane, highly permeable to O_2 and not permeable to glucose on one side, and a crosslinked protein gel permeable to both oxygen and glucose on the other side [20, 90, 92]. Further, to address the O_2 limitation for glucose biosensors, Wang et al., developed carbon paste based GOx electrodes that can accurately measure glucose even in O_2 deficient solutions. They achieved this by incorporating fluorocarbon (Kel-F oil) pasting liquid or more recently PDMS oils, within the carbon paste-GOx layer, making them permeable to O_2 fluxes up to 45 to 50 fold higher than in water [88, 89].

Such sensors achieved a linear glucose detection limit up to 20 mM and allowed assays as high as 40 mM [93].

Optical methods for measuring GOx mediated O_2 depletion were also proposed, wherein GOx was immobilized on optical fibre in conjunction with a dynamic quenching fluorophore, e.g., tris(1,10-phenantroline)-ruthenium(II) [94–96]. However, such luminescence oxygen optrodes measure not only the O_2 depletion due to GOx catalyzed reaction, but also the physiological changes in O_2 concentrations. To overcome this problem, Li et al., proposed a dual fibre-optic fluorescent sensor [97]. The difference between O_2 concentration measurements by sensors with and without GOx was used to correlate glucose concentrations.

4.3.2 Measuring H_2O_2 production

The electroactive H_2O_2, produced by the O_2 mediated electrooxidation of glucose using GOx, can be assayed by three methods: First, the catalytic electrooxidation of H_2O_2 at traditional electrodes between +0.3 to 0.8 V *vs.* saturated calomel (SCE) or Ag/AgCl electrode; second through the catalytic electroreduction between +0.0 to −0.2 V *vs.* SCE; or third through H_2O_2-oxidation of a peroxidase, usually HRP, followed by mediated or direct electron transfer from the oxidized peroxidase to the electrode [1]. The delays due to additional reactions to transfer electrons from GOx or other redox enzymes to the electrodes, the high over-voltages for amperometric anodic measurement of H_2O_2 and the relatively slow temporal resolution are the main limitations for measuring H_2O_2 and in general, for indirect enzymatic electrochemical sensors. To circumvent these limitations, nanomaterials are incorporated in electrodes for increasing the electron transfer kinetics, sensitivity and response time, while reducing the over-voltages [1, 17, 98].

One of the objectives for incorporating nanomaterials in glucose sensing electrodes is to tune the operating over-voltages to optimal +0.3 to +0.5 V for anodic electrooxidation and +0.0 to −0.2 V for cathodic electroreduction of H_2O_2 [99, 100]. The optimal over-voltages aid in efficient and preferential detection of H_2O_2 through electrooxidation or reduction that respectively lower or eliminate the contributions from the easily oxidizable common electrochemical interferents during electrooxidation or electroreduction respectively. However, it must be emphasised that different nanomaterials help improve different sensor properties (e.g., sensitivity, response time, and/or over-voltages), and often combinatorial approaches are needed to tailor sensor responses.

A simple approach again is the use of nanoporous platinum electrodes. Wang et al. electro-deposited Pt nanowires or Pt black surfaces of varying roughness on Pt electrode, loaded up to 3-fold higher GOx content (than on smooth non-porous Pt electrode) by electrochemical co-deposition of GOx entrapped in a conducting polypyrrole film, and demonstrated about 150-fold increase in sensor sensitivity [101]. Although this sensor required high over-voltage (+0.7 V), interference from ascorbic acid and dopamine was significantly low due to the polypyrrole film.

Nano-materials based on metals, their alloys or oxides, as described in Section 4.1, are capable of lowering the over-voltages for electrochemical reactions. However, they need to be immobilized at the electrode surface and some sort of electrical wiring is required for the efficient transfer of electrons from the nano-materials to the electrode. Commonly, the nano-materials and GOx are co-incorporated on electrodes through physical adsorption, electrochemical deposition, or immobilization using polymers or ceramics (composites). Elzanowska et al. provided a brief review of the various NPs based on metals, alloys and oxides co-deposited with GOx on a variety of metal- and carbon-based electrodes for detection of H_2O_2 [102]. Often the same electrode can measure cathodic and anodic currents from the redox reactions of H_2O_2, e.g., a sensor with Ir oxide and GOx co-deposited on electrodes can measure the oxidation of H_2O_2 at about +0.4 V, while the reduction at −0.1 V [102–104]. Yet, it is the electroreduction of H_2O_2 that is preferred for glucose sensing, because, its low operating voltages between −0.2 and 0.0 V not only eliminates interference from common electroactive substances, but also maintains very low background currents in complex biological fluids.

Horseradish peroxidase (HRP) is a natural enzyme that oxidises H_2O_2 and the transfer of electrons from reduced HRP to the electrode occurs between −0.1 and +0.2 V. The over voltages depend on the nature of the electrical wiring of the redox centre of HRP to the electrode. GOx and HRP entrapped in an electro-polymerized polypyrrole film needed over-voltages around +0.15 V [105, 106]. Further, when Ohara et al. separated the electrically wired HRP (directly on electrode surface) from the (outer) GOx layer, the over-voltages typically were at about 0.0 V [107]. Similar results were also observed when the electron transfer from redox centre of HRP to electrode is mediated with artificial mediators such as ferrocene and metalized polymers [108–114]. However, long-term stability of HRP at physiological temperature is a concern. Therefore, thermostable peroxidases from other sources including soybean have also been tested [115–118].

Metal-hexacyanoferrate electrocatalysts, especially Pursian blue (PB, ferric-ferrocyanide), are another group of nanomaterials that require very low over-voltages for electrocatalysis of H_2O_2 redox reactions [119]. PB, often regarded as an artificial peroxidase, is shown to require an over-voltage of −0.1V vs. Ag/AgCl for H_2O_2 detection [120, 121]. Screen-printed miniature sensors for glucose and H_2O_2 detection were also developed using PB-carbon inks mixtures [122].

Electroreduction at metalized carbons on common electrodes also demonstrate preferential detection of H_2O_2 at low over-voltages. Metal and alloy particles of Ru-, Rh-, Ir-, or Pt-Ru are dispersed in carbon-paste enzyme electrodes to detect H_2O_2 at about 0.0 V having high selectivity and sensitivities

[99, 100, 123]. Compact graphite modified with Pt and Pd NPs where shown to detect glucose at −0.05 V *vs.* Ag/AgCl, having a detection limit of 10 μM and a sensitivity of 0.109 μA/μM [124]. Co-hexacyanoferrate NPs, chitosan and CNT modified GCE electrode demonstrated an interference free detection of glucose at −0.2 V, having a detection limit of 5 μM and a linear range of 0.01 to 10 mM [125]. Hrapovic et al. co-immobilized Pt-NPs, CNTs and GOx in a Nafion film and showed higher sensitivities than when the Pt-NPs or CNTs were incorporated alone [126]. CNTs on their own also showed low H_2O_2 electro-reductive over-voltage of −0.05 V *vs.* Ag/AgCl when co-immobilized with GOx in Nafion [127].

Electrode coatings combining metal NPs with other polymers were also used for H_2O_2 electroreduction based glucose sensors. Au NPs in silicate network on tin oxide (ITO) glass electrode showed a linear glucose detection range of 0 to 20 mM at the reductive potential of −0.2 V [128]. Gu et al. immobilized Cu^{2+} electrocatalytic ions in a poly-ion complex of DNA and poly(allylamine) on GCE to measure H_2O_2 at −0.2 V *vs.* Ag/AgCl [129].

It must be emphasized that electrode coatings combining metal NPs and carbon materials often provide higher sensitivity for electrochemical detections than when the NPs or carbon materials are used alone. However, concerns are raised regarding the higher costs and low reproducibility in the production of metal NPs- and CNTs-based electrode coatings.

Nanomaterials can also be explored as cost-effective alternatives for conventional precious-metal electrode based glucose biosensors. Recently, Zhu et al. demonstrated that a 10-fold smaller-sized CNT-fibre based electrode can perform similar to or better than the traditional Pt-Ir electrode in the amperometric sensing of glucose [130]. In this approach, the exceptional properties of CNTs were consolidated into a macro-CNT fibre prepared by chemical vapour deposition and this inexpensive fibre was explored as an alternative for the expensive Pt-Ir.

An optical method for indirect measurement of glucose through H_2O_2 reaction with a bis(2,4,6-trichrorophenyl) oxalate (TCPO) to form a peroxyoxylate fluorophore, was proposed by Abdel-Latif et al. [131]. The glucose concentration dependent increase in fluorophore is detected optically.

4.3.3   Measuring pH changes

Gluconic acid is a product of the GOx and GDH catalyzed oxidation of glucose. Conventionally, the hydrogen ions that are generated by the dissociation of gluconic acid to gluconate are detected and related to glucose concentration. However, due to the low dissociation constant for gluconic acid, the depleting O_2 concentration, and the buffering capacity of physiological fluids, the glucose sensors suffer from low sensitivities, narrow dynamic detection range, longer response time (10 to 20 minutes) and inherent drift in pH measuring signal (requiring frequent sensor calibration). Seo et al. proposed the electrooxidation of H_2O_2 at about +0.7 V in the vicinity of an ion selective field effect transistor (ISFET) sensor to generate further $2H^+$ ions, while regenerating an O_2 molecule thereby increasing the sensor sensitivity and extending the dynamic detection range for the sensor [132]. They observed four fold improvement in sensitivity, extension of dynamic range to 5 mM and a faster response time of 3 min in 10 mM PBS (pH 7.4, 100 mM NaCl). Sensor drift of ISFET sensors typically occurs due to accumulation of H^+ ions at the electrode. To lower the sensor drift, Park et al. proposed an ISFET sensor coupled with electroreduction of H_2O_2 at -0.7 V platinum electrode that generates OH^- ions, which diminish the H^+ ions leading to faster regeneration of the enzyme reaction [133]. Through this method sensor sensitivity drift is lowered to less than 5%, response time to about 2 min and a long-term stability observed for 15 days.

The nanotechnology, behind the design of novel FETs (semi-conductor chips often based on silicone materials), inherently miniaturize the sensor, while improving the sensitivities and detection ranges for glucose sensing. Through further miniaturization of the ISFET sensor and its better electrode configuration (based on dissociation of gluconic acid and electrooxidation of H_2O_2), Lee et al. achieved a response time of about 1 min [134]. The ultimate miniaturization of such sensors was demonstrated by Besteman et al. who designed 1-dimensional CNT nano-materials for conductivity based glucose nanosensors. GOx enzyme was attached on the side wall of individual CNT through a linking molecule and the enzyme coated CNTs functioned as pH sensors that showed large and reversible changes in conductance up on changes in pH. Increase in glucose concentration induced increase in conductance and that too at the level of a single CNT molecule [135]. The advantage of molecular-scale nanosensors in conductivity measurements was further demonstrated by Forzani et al. who developed a glucose nano-sensor based on conducting-polymer-based nano-gap [136]. Two nano-electrodes separated by 20 to 60 nm are bridged by a polyaniline/GOx conducting nano-junction. Changes in conductance of the nano-junction in relation to glucose oxidation are detected. The small size of the nano-junction sensor is attributed to cause natural regeneration of the enzyme, while consuming minimal amount of O_2, resulting in a very fast response time (<200 ms) [136]. The use of nano-materials in the design of FETs is being widely researched and the discussion reported in the study of Lee et al. elucidates the relation between sensor design and pH/glucose sensing mechanism [137].

4.3.4   Enzymatic optical glucose sensing

The unique optical properties of quantum dots and metal NPs were combined with the specificity of enzymes to develop reversible sensing probes for glucose biosensing. Bahshi et al. showed that CdSe/ZnS QDs or Au-NPs functionalised with methylene blue (MB) when coupled with glucose dehydrogenase generates measurable optical signal for glucose sensing [138]. The interactions between biocatalytically generated H_2O_2 using GOx and CdSe/ZnS QDs also enable ratiometric fluorescence/luminescence analysis for glucose detection [139–141]. Duong et al. conjugated GOx and HRP enzymes on CdSe/ZnS QDs to design a Cd-FRET probe for glucose detection having a linear detection range of 0 to 28 mM [142]. Furthermore, an oxygen sensitive photobioelectrochemical sensor using CdSe/ZnS nanocrystals and GOx was also developed for potentiostatic or potentiodynamic glucose measurements [143]. Wu et al. proposed a phosphorescent detection mode that avoids the interference from autofluorescence and scattering of light from biological samples. They developed Mn-doped ZnS quantum dots conjugated with GOx for phosphorescent sensing of glucose [144].

Towards the development of optical CGM devices, Kim et al. developed a platform technology for the encapsulation CdSe/ZnS QDs and GOx enzyme within an Fmoc-diphenylalanine self-assembled peptide hydrogel [145]. The efficacy of the resulting photolumeniscent hydrogels for glucose sensing was demonstrated with minimal leakage of QDs or GOx due to their efficient immobilization within the fibrous hydrogel network. Similarly, Liu et al. combined CdTe, CNTs, GOx and Nafion to obtain composites that showed better sensitivity for glucose than that based on GC electrode modified by CdTe QDs or CNTs alone [146].

4.4 Direct enzymatic glucose sensing

The redox centres of the enzymes, after accepting electrons from glucose, must reach the close proximity (less than about 3–5 Å) to the electrode surface to produce any meaningful currents for measurement. However, such close proximity is difficult to achieve because the enzymes developed insulating protein or glycoprotein shells that protect the redox centres from indiscriminate electron exchange with other co-existing redox macromolecules [22]. For instance, the $FADH_2$ (redox centre) of GOx is reported to be buried at a depth of about 13–15 Å, at which electron tunnelling distance, due the exponential decrease in electron-transfer rates vs increasing electron tunnelling distance, electron transfer to an electrode is negligible [1, 16, 25]. Hence, to achieve a measurable electron transfer from the redox centre of GOx to electrode, mediation through an additional redox couple is essential. The O_2/H_2O_2 redox couple is considered the natural mediator for GOx that forms the firm basis for first generation glucose sensors [1]. Due to the low solubility of O_2 (~0.2 mM) in physiological solutions, strategies to regulate glucose/O_2 concentration ratios reaching the sensor have been developed. On the other hand, to altogether avoid O_2 dependence, the O_2/H_2O_2 redox couple is replaced with artificial mediators in the second generation glucose sensors, thus achieving direct amperometric sensing based on shuttling of electrons from enzyme's redox centre to electrodes.

4.4.1 Diffusionally mediated amperometric biosensors

In a typical second generation glucose sensor, a mediator collects electrons from the enzyme, diffuses to and delivers electrons at the electrode surface. Such diffusionally mediated amperometric biosensors require their mediators to react rapidly with reduced enzyme at low over-voltages, have stable oxidized and reduced forms, and be independent of pH and O_2 partial pressure changes [28]. Since the introduction of ferrocene carboxylate/ ferrocinium carboxylate couple as an effective redox couple for GOx by Cass et al. [147], a wide range of artificial mediators, as reviewed by Heller and Feldman, and Chaubey and Malhotra [1, 28], have been developed and investigated for direct amperometric glucose sensing (Table 4-1).

The redox potentials of the mediators are usually close to that of the redox centres of the enzymes, thus their redox reactions, often, do not require high over-voltages. Several organic dyes and inorganic redox ions are reported to have poor stability and susceptibility to pH changes. Ferrocene and their derivatives are the most widely researched mediators, among which 1,1'- dimethyl-3-(2-amino-1-hydroxyethyl) is commercially used for MediSense (now part of Abbott) blood-glucose meters. $Fe(CN)_6^{3-/4-}$ redox couple is also used in home blood-glucose monitoring strips [1]. Other, promising mediators include quinone, phenothiazine and phenoxazine compounds, as well as tetrathiafulvalene (TTF), tetracyanoquinodimethane (TCQM), TTF-TCQM conducting salts [1, 17, 28].

Recent research has also shown CNTs as mediators for direct electron transfer from redox centre of enzyme to the electrode surface. GOx is reported to spontaneously adsorb on CNTs, while maintaining its glucose-specific enzyme activity [148, 149]. Guiseppi-Elie et al., observed the tubular SWNT fibrils to position themselves within tunnelling distance of the cofactors, with little denaturation to the enzyme's glycoprotein coating, facilitating a quasi-reversible one-electron transfer directly from GOx-$FADH_2$ to electrode (GCE and free-standing SWNT papers) [148]. On the other hand, Zhao et al. reported the adsorption of GOx on MWCNTs caused conformational changes on glycoprotein coat of GOx (without denaturation) leading to a less shielded active site that facilitates direct electron transfer [149]. Ivnitski et al. demonstrated high electron-transfer rate constants for both MWNTs and SWNTs [15, 150]. Yet due to the superior electron-transfer properties of SWNTs over MWNTs, they later incorporated SWNTs and GOx in a bioinorganic silica composite. Within this electron conductive composite, SWNTs work as nanowires that increase the surface area for enzyme immobilization. The carbon electrodes, modified with GOx-SWNT-silica composites, had a formal redox potential of −406 mV at pH 6.2 vs. Ag/AgCl, close to that of the FAD/$FADH_2$ redox centre of GOx, and facilitated direct electron transfer to the electrode [150]. Similar results were also reported, where in, electrodes such as glassy carbon

Table 4-1 Artificial mediators – redox couples for electron shuttling from redox centres of enzymes such as GOx-FADH$_2$ and GDH-PQQH$_2$ to electrodes [1, 28, 157]

Type	Mediators
Organic	Quinoid dyes: methylene blue, thionine, pyocyanine, Safranine T, brilliant cresyl blue, azure A
	Quinone and its derivatives including polymeric quinones: hydroquinone, benzoquinone, 1,10-phenanthroline quinine, polyaminonapthoquinone (PANQ)
	Viologen derivatives: N,N'-Di(4-nitrobenzyl)viologen dichloride, poly(o-xylylviologen dibromide), poly(p-xylylviologen dibromide)
	Phenothiazines, Phenoxazines, Dithia- & tetrathia-aromatic compounds
	Phenazines: N-ethyl phenazine, N-methyl phenazine
	Wurster's salt: N,N,N',N'-Tetramethyl-p-Phenylenediamine (TMPD)
	Heterocyclic dihydropolyazines: 5,10-dihydro-5,10-dimethylphenazine, 1,4-dihydro-1,3,4,6-tetraphenyl-s-tetrazine
	Poly(1-(5-aminonapthyl) ethanoic acid) (PANEA)
	Tetrathiafulvalene (TTF)
	Tetracyanoquinodimethane (TCQM)
	Conducting salts: TTF-TCQM, N-methyl phenazinium-TCQM (NMP-TCQM)
	Carbon materials: Carbon nanotubes (CNTs), powdered graphene, fullerenes
Inorganic	O$_2$/H$_2$O$_2$
	Metal hexacyano-complexes: Fe(CN)$_6$$^{3-/4-}$, Co(CN)$_6$$^{3-/4-}$, Ru(CN)$_6$$^{3-/4-}$
Metal-Organic	Pentacyanoferrate(III)-, Fe$^{2+/3+}$-, Ru$^{2+/3+}$-, or Os$^{2+/3+}$-complexes with organic polymers such as pyridine, pyrazole, imidazole, histidine, aza- and thia-heterocycles, benzotriazole, benzimidazole, and aminothiazole
	Ferrocene and its derivatives: Ferrocene, 1,1-dimethyl ferrocene, vinyl ferrocene, ferrocene carboxylic acid, ferrocene metanol, hydroxymethyl ferrocene, 1,1'-dimethyl 1,3-(2-amino-1-hydroxyethyl) ferrocene
	Nikelocene
	Manganese cyclopentadienyl (Cp) half-sandwich, chromium half-sandwich

or graphite were modified with CNTs or graphene co-immobilized with GOx in materials such as ionic liquid reconstituted cellulose, boron-doped silica sol-gel, chitosan, Nafion and chitosan, and sometimes in combination with other mediators such as prussian blue (hexacyanoferrate(III)) and ferrocene carboxylic acid [151–155]. In addition to CNTs, nano-particles of Au, Pt and laponite colloids (polycrystalline materials made of hydrous sodium lithium magnesium silicate) have also been reported for direct electron transfer from GOx-FADH$_2$ to electrode [153, 154, 156]. The colloidal nano-particles have dimensions (e. g., AuNPs have a particle size of 13 Å) that allow them to reach within the electron tunnelling distance of GOx-FADH$_2$ (<3 Å [22]).

The freely diffusing mediators of nano-dimensions are highly toxic to living systems, and hence are not suitable for implantable CGM systems. But, for measuring glucose in withdrawn samples of blood, outside the body, the artificial mediators have widely replaced the natural O$_2$/H$_2$O$_2$ redox couple in home glucose monitoring-strips and in lab-based analytical systems [18, 23].

An optical means of direct enzymatic detection was also reported using the redox mediator tetrathiafulvalene (TTF) [158]. Oxidized TTF absorbs light in the 540–580 nm range. It becomes reduced by receiving electrons from GOx-FADH$_2$, thus lowering the light absorption.

4.4.2 Electrical wiring of enzyme's redox centres to the electrode

Direct electrical wiring between enzyme's redox centre and electrode can further increase the current densities from glucose redox reactions reaching the electrode (including conventional smooth and non-porous electrode). Heller's group developed efficient non-diffusional routes to covalently connect the GOx-FADH$_2$ to the electrode through redox polymers and achieved current densities above 1 mA/cm^2 at 0.04 V vs. Ag/AgCl in their commercially available second-generation CGM system [1, 17, 18, 22, 23]. The huge current densities allow the design of ultra-small electrodes. The small size, high sensitivity, low over-voltages (0.0 to 0.1 V), negligible interferences, and by avoiding the leaching of diffusional mediators, electrically wired systems are particularly attractive for implantable CGM [17, 18].

Evolutionarily, redox enzymes developed sophisticated protective protein/glycoprotein shells, due to which, irrespective of the applied over voltages, their active redox centres are neither oxidized nor reduced directly at the electrode surface. To facilitate the direct electrical wiring, chemical alterations to the protein/glycoprotein layers were tested by covalent binding of synthetic redox mediators on the enzyme's protein shell. Hill first described the design of electron relays modified enzymes by covalently attaching ferrocene carboxylic acid to lysine residues of GOx using isobutyl chloroformate [159]. Later, Heller's group also covalently attached ferrocene carboxylic acid to GOx, but through carbodiimide chemistry, while partially unfolding the enzyme protein by 2–3 M urea treatment [22–25]. Depending on the number of artificial mediators covalently attached per enzyme molecule and the degree of denaturation of the protein shell, the modified GOx communicated electrically with electrodes. The covalently attached mediators functioned as electron relays, enhancing the electron transfer distances between 20 to 25 Å compared to <3 Å for the native enzyme [22]. However, the freely diffusing relay-modified enzymes were not efficient enough due to random molecular orientations on electrode, competition from O_2 and reliance on containing membranes similar to that with artificial mediators.

When the electron-relays on modified enzymes were covalently bound to a redox polymer network, a segment of which was bound to the electrode, irrespective of the molecular orientation, the immobilized enzymes were electrically connected to the electrode. This method also allowed multiple relay-modified enzyme layers, enhancing the current densities to mA/cm^2. The redox hydrogels utilized for immobilization of relay-modified enzymes include poly(vinylpyridine) and poly(vinylimidazole) containing covalently linked Os-redox complexes with pyridine and imidazole derivatives [1].

Covalent binding of GOx on nanomaterials such as Au-NPs and CNTs or Graphene is also being tested for direct electrical wiring. Willner et al. covalently attached GOx on single Au-NPs and assembled them on Au electrodes leading to an effective electrically contacted enzyme electrode [160]. Willner's group also demonstrated the covalent binding of reconstituted GOx on SWCNTs containing end-plane carboxylate functionalization through carbodiimide coupling. The SWCNTs were first covalently attached, vertically aligned on Au electrode and then their free-ends coupled with reconstituted GOx. The SWCNTs functioned as conductive nanoneedles connecting the enzyme's redox centre to the electrode, thus achieving electron transfer distances higher than 150 nm [161]. Gooding's group also reported aligned SWCNT arrays for efficient electrical wiring of native and reconstituted GOx [162]. In the reconstituted GOx model, FAD cofactor is first covalently attached on the NP or CNT, and then the apo-enzyme assembled on the covalently attached FAD cofactor. This method yields aligned enzymes on electrode surface, resulting in efficient electron transfer from each attached enzyme. The reconstituted GOx assembly on Au-NPs and CNTs described above has the limitation of assembly of mono-layers of enzymes that leads to low overall current densities compared to the high densities observed with enzymes immobilized in hydrogels developed by Heller's group. Recently, Willner's group reported new method to fabricate crosslinked three-dimensional metallic NP/enzyme composites [26, 27]. The three-dimensional conductivity of the Au NPs composite contain redox active bridging bisaniline units, acting as relays, facilitated the electron transfer between the redox sites of multiple enzyme layers and the electrode. The resulting sensors are shown to achieve amperometric responses to glucose concentrations ranging from 2 to 120 mM, and levelling off above 120 mM at 0.3 V *vs.* SCE [27]. Other directly wired electrodes include composite coatings based on GOx covalently bonded to DNA on graphene-AuNP hybrid electrodes [163] providing a glucose sensor with lower detection limit (0.3 µM), high sensitivity (24 mA mM^{-1} cm^{-2}) and long-term stability (80% of initial response after 4 months) compared to other graphene-AuNP electrode based sensors [164, 165].

5. Smart membranes for glucose sensing

Nano-structured membranes for various applications including biosensing have been comprehensively reviewed elsewhere [166–168]. The membrane materials, their structure and porosity play an important role in their selective permeability to solutes. They are said to be nano-porous when their pore sizes are less than 100 nm, and sometimes the upper limit is extended to 500 nm. Further, based on porosity they are classified into four categories: 1) non-porous, 2) microporous (≤2 nm), 3) mesoporous (2–50 nm), and 4) macroporous (50–500 nm) [167]. Membranes based on all four categories have been used for glucose biosensing and their use has four major purposes: 1) to entrap or immobilize enzyme or other glucose affinity molecules at sensing surface, 2) to avoid interferents (that compromise sensor function) from reaching the sensing element, 3) to widen the dynamic glucose measuring range (covering the required 2–30 mM), and 4) to tailor the foreign body responses to implantable glucose biosensor and enhance their reliable sensing lifetime upon implantation.

5.1 Membranes to entrap or immobilize enzyme or affinity molecules

The first glucose enzyme electrode proposed by Clark and Lyons advocated the sandwiching of GOx enzyme between an oxygen electrode and a semi-permeable dialysis membrane. Considering the hydrodynamic diameter of about 100 Å of GOx, entrapping GOx within a dialysis membrane would be difficult to achieve for long-term glucose sensing. Later, Updike and Hick immobilized GOx within a polyacrylamide gel that covers an O_2 electrode for rapid glucose sensing [169]. Thereafter, a wide variety of materials, often in combination with mediators and covalent immobilization of GOx have been developed for immobilization of GOx on sensing elements.

The most common among the methods for immobilization is to crosslink GOx on proteins, especially bovine serum albumin, using glutaraldehyde [170]. This method can immobilize large quantities of GOx in multiple layers, generating large quantities of H_2O_2 that diffuse to electrode surface for detection. Direct electrical wiring of multiple layers of enzymes, through electron relays and redox hydrogels, is another method that provides highly efficient enzyme immobilization for the direct transfer of electrons from the enzyme's redox centre (e.g., GOx-FADH$_2$) to the electrode surface (discussed in Section 4.4.2). Such methods achieved the highest current densities for amperometric glucose detection. Monolayers of covalently attached enzymes on electrodes such as thiolated Au electrodes also showed efficient transfer of electrons, but the currents generated are often too small. But when the enzyme mono-layers were covalently attached on CNTs or ISFETs, the resulting sensors were highly sensitive and miniaturized (details in Sections 4.3.3 and 4.4.2).

Electro-polymerization is one method commonly used to immobilize oxidizing enzymes for glucose. Metal oxides such as TiO_2, ZnO, and CeO_2;

and SiO_2 are used to generate nanoporous thin films entrapping enzymes on electrode surface [171–175]. In addition to metal and silicon oxides, GOx is also co-electrodeposited with polymers, including polyphenylenediamine, polyphenols, polyaniline, polypyrrole, and polythiophene [176–178]. Sol-gel methods are also used to entrap GOx in materials such as silica, titania or CeO_2 [179–182]. The advantage of the sol-gel method is that it works at room temperature, and does not affect the GOx activity.

Layer-by-layer (LBL) self-assembly, a widely explored method for immobilization of GOx and GDH on electrodes, provides precise control in the composition, thickness and nanostructure of the resulting membranes [183]. Sequential adsorption of oppositely charged polyelectrolytes has been the primary approach for LBL assembly and other molecular interactions explored include hydrogen bonding, hydrophobic and biological interactions. Enzymes being proteins have an overall charge, which property is exploited in LBL assembly. For example, GOx has an overall negative surface charge, resulting in it forming polyion complexes with polycations. Polycations commonly used for GOx immobilization include, poly(allylamine), poly(L-lysine), poly(ethylenimine), poly(dimethyldiallylammonium chloride), poly(allylamine hydrochloride), poly(pyridine), poly(aniline) and chitosan [183]. Often, the polycations are functionalized with either metals ions such as osmium, or mediators such as ferrocene, allowing the polymers to function as electron relays [184–187]. Nanoparticles, depending on the method of synthesis, contain charges, facilitating electrostatic adsorption. The NPs tested include, MnO_2, SiO_2, TiO_2, Pt, clay and functionalized quantum dots, and the stability of NP/protein multilayers was said to be better than that of polyelectrolyte/protein films [183]. CNTs have also been assembled by LBL method to immobilize enzymes. Liu et al. incorporated NADH in polyaniline and poly(aminobenzenesulfonic acid) modified SWCNTs LBL assemblies and demonstrated the electrocatalytic ability of NADH, showing great promise for use of NADH based dehydrogenases for biosensing [188]. Biotin labelled GOx was assembled alternating with avidin molecules to obtain catalytically active multilayer films for glucose sensing [189]. Similarly, anti-IgG labelled GOx was assembled by LBL assembly with IgG on GCEs coated with gelatin [190]. In spite of the precise control of the membrane structure, the long-term stability of enzymes in the LBL complexes is poor and the enzyme stability is reported to improve when the LBL assemblies are crosslinked. For example, LBL thin films based on GOx and poly[vinylpyridine $Os(bisbipyridine)_2Cl$]-co-allyamine were crosslinked using glutaraldehyde vapours, crosslinked films were obtained, which retained nearly 100% enzyme activity for three weeks [191].

GOx is also immobilized on electrodes through its co-electrospinning with polymers or by covalent attachment on electrospun nanofibrous mats [192]. Ren et al. developed an electrospun poly (vinyl alcohol) (PVA) membrane to immobilize GOx on the surface of an Au electrode for an amperometric biosensor [193]. Such sensors showed rapid response time of ~1 s, linear response range from 1 to 10 mM and detection limit of 0.05 mM [194]. Manesh et al. fabricated nanofibrous glucose electrodes by electrospinning polymethylmethacrylate, GOx and MWCNTs wrapped by a cationic polymer (poly(dimethylammonium chloride)) that had a response time of ~4 s, linear response range of 20 μM to 15 mM and detection limit 1 μM [195]. Using a similar technique, GOx was immobilised through electrospinning using DNA/SWCNTs/ poly(ethylene oxide) that extended the linear response range to 20 mM, with sensitivity of 2.4 mA cm^{-2} M^{-1} [196]. Besides the semi conducting and insulating materials described above, conducting polymers were also used to fabricate electrical sensors. The eletrospun polyaniline (PANi)/PS nanofibers encapsulated GOx had higher sensitivity to glucose [197]. The electrospun systems offer large surface area, high porosity and interconnectivity, nanofibrous structure ensuring high enzyme immobilization that improve sensor sensitivity, while shortening the sensor response time.

Enzymes are also entrapped within polymer films such as Nafion, chitosan and cellulose acetate, often combining multiple polymers to form composites. In addition, following the successful demonstration of ferrocene mediated direct transfer of electrons from GOx-FADH$_2$ redox centre to electrode [147], wide variety of mediating materials have been investigated to design second generation sensors. The materials include powdered graphite (carbon paste), CNTs, CNT fibres, fullerenes, nano-metals, nano-metal arrays, electrocatalysts, nanowires, metal and metal oxide NPs, organic/ inorganic electro-catalysis mediators have been co-immobilized using binders [157]. Polymers such as Nafion, chitosan, polymethyl methacrylates, polycarbonates, polyvinyl chloride/alcohol; various polyelectrolytes, Ionic liquids, and other organic materials have been used as binders [157].

5.2 Membranes for diminishing interferents

Measuring glucose in complex biological fluids such as blood and interstitium has been a challenge. A common problem is the multiple competing interferents. Sugars similar to glucose such as fructose, maltose, xylose, and galactose compete for sensing reactions. Typically, exclusive selectivity to glucose in the presence of other competing sugars is achieved by the use of GOx enzyme. Another problem with most electrochemical glucose sensors is interference from electroactive species in biological fluids including ascorbic acid, uric acid, and acetaminophen. Most co-existing electroactive interferents oxidise at over potentials between 0.2 and 0.8 V, causing selectivity problems for electrochemical biosensors. Hence, a major objective for electrode design has been to tune over voltages to between −0.2 to +0.2 V for preferential oxidation or reduction of glucose, H$_2$O$_2$ or artificial mediators at the electrode, wherein the contributions from co-existing interferents are eliminated or negligible (described in Sections 4.3.2 and 4.4). However, most conventional electrodes require high over voltages, usually around +0.6 V, necessitating the need for selective membranes to diminish competing interferents.

Glucose does not spontaneously oxidize in the presence of strong oxidants such as PbO_2, BaO_2, CeO_2 and MnO_2, while co-existing electroactive interferents do. Thus, introducing a pre-oxidizing layer based on a strong oxidant exterior to the enzyme layer oxidizes interferents to electrochemically inactive forms before they can reach the electrode surface [198]. The unaffected glucose (analyte) passes through the pre-oxidizing layer to react with the enzyme generating electroactive species (primarily H_2O_2) used to accurately estimate glucose concentrations. However, the ability of the pre-oxidizing layer in eliminating just interferents is dependent on the nature of its deposition. For example, when metal oxide powders were mixed with polymers and applied manually on enzyme electrodes, the metal oxide NPs diffuse into enzyme layer and compromise sensor function by decomposing the enzymatically generated H_2O_2 [199]. To avoid direct contact with the enzyme layer, Choi et al. used an additional permselective membrane sandwiched between the enzyme and pre-oxidizing layers. They also showed that pre-oxidizing layers made with PbO_2 NPs yielded biosensors with the best performance compared to BaO_2, CeO_2 and MnO_2 [199]. Earlier, disposable glucose biosensors made by screen printed thick-films of PbO_2 and GOx electrodes on nitrocellulose paper were prepared such that analyte samples travelled chromatographically through PbO_2 pre-oxidizing layer before the analyte (glucose) reaches the GOx layer [200]. The resulting sensors were shown to be virtually non-susceptible to interferents including ascorbic acid, uric acid and acetaminophen. Later, Xu et al. electrodeposited MnO_2 NPs dispersed in chitosan to form a stable interference eliminating oxidant membrane on enzyme electrodes [201].

The most popular among the interference eliminating methods is the use of permselective membranes. They, typically, exclude interferents based on size, charge or polarity. Size exclusion is primarily dependent on the pore size and distribution within the membrane. Uniform pore size is commonly achieved through electrodeposition using materials such as poly(phenylenediamine), polyphenol, poly(aminophenol), substituted naphthalenes, 3-amino-propyl-trimethoxysilane and poly(1,2-diaminobenzene) [176, 202–208]. The advantage of the electrodeposition method is that miniature sensors can be coated with uniform, thin and compact films [198]. Furthermore, plasma polymerization method is also used to deposit size exclusion membranes based on cellulose acetate and mercaptosilane on sensors [209–211]. Electrostatic repulsion using membranes containing charged functional groups, especially a negatively-charged perfluorinated ionomer – Nafion, is also widely studied for elimination of interferents [212]. In addition, hydrophobic membranes made of alkanethiol or lipid layers are used to exclude hydrophilic interferents [213]. Often, different membranes are used in combinations to eliminate multiple interferents in complex biological fluids. For instance, Moussy et al. sandwiched GOx immobilization layer between an electro-deposited poly(phenylenediamine) inner layer and a Nafion outer layer [212]. Combinations of Nafion and cellulose acetate have been used to eliminate the interference of the neutral acetaminophen and negatively charged ascorbic and uric acids, respectively [214, 215]. Rodriguez et al. prepared a layer-by-layer assembly of polyethylenimine, Nafion, polyethylenimine, and DNA followed by alternate deposition of polyethylenimine and GOD, on thiolated gold surface, considerably improving the selectivity and sensitivity of the glucose biosensor [216].

5.3 Widening the dynamic (linear) measuring range

The mass transport limiting membranes (also referred as semi-permeable or flux-limiting membranes) are widely used for controlling the diffusion of substrate (e.g., glucose) to the sensing element. The primary purpose of these membranes is to allow sensors to function in physiological glucose concentration range by controlling solute transport, such that glucose flux reaching the sensing element is the rate limiting step for assay. This allows measurements on un-diluted biological fluids, making implantable CGM a possibility [217, 218]. Secondly, the membranes also slow down, if not prevent, the interferents from reaching the electrode. In the case of metal electrodes, especially platinum, they also prevent electrode poisoning. Often, the permselective membranes, such as Nafion and cellulose acetate (described Section 5.2), in addition to eliminating certain types of interferants, also function as mass transport limiting membranes. A third function for the membranes is to protect underlying NPs (metal, metal oxides, carbon materials, fluorescent molecules, or lectins), electro-catalysis mediators, LBL enzyme assemblies, or electrodeposited films from degradation and leaching. Furthermore, for implantable sensor applications, it is essential to covalently immobilize the NPs, redox mediators and enzymes to sustain reliable sensor function and to minimize their toxicity on surrounding tissues.

It must be emphasized that any material deposited on electrode surface introduces delay in the transport of the analyte, its co-reactants and reaction products, affecting the sensor function. Thus, even the enzyme layer (described in Section 5.1) functions as mass transport limiting membrane. However, the direct access of the enzyme molecules to analyte solution causes rapid saturation leading to lower dynamic detection ranges. To cater for the clinical need of sensors measuring in the range of 2 to 30 mM for diabetic patients, mass transport limiting membranes are essential for most sensor designs [4, 217]. Thicker the membrane, the slower will be the analyte transport. Hence for best results, the membranes should be the thinnest and their trans-membrane porosity as uniform as possible [166, 219, 220].

Polymers and co-polymers tested as mass transport limiting membranes include cellulose acetate (CA) [221, 222], poly(bisphenol A carbonate)

(PC) [223], perflurocarbon [224], poly(vinyl chloride) (PVC) [225], poly(tetrafluoroethylene-co-vinylidene fluoride-copropylene) [226], poly(4-vinylpyridine-co-styrene), Nafion [227, 228], polydimethylsiloxane [221], polyallylamine-polyaziridine, polytetrafluoroethylene [229], poly(ethylene oxide), and polyurethane (PU) [170, 230]. Composite membranes e.g., CA-PC-PVC-PU and poly(hydroxyethyl methacrylate)-poly(dihydroxypropyl methacrylate)-N-Vinyl pyrrolidone-ethylene glycol dimethacrylate (HEMA-DHPMA-NVP-EGDMA) were also tested [170, 231]. Among the materials, PU membranes have been the most popular due to their durability, biocompatibility and long-term stability of sensor function *in vitro* [170, 232]. They are primarily deposited on sensors through solvent casting. Often additives such as epoxy resins are used to reinforce the polymers for improving mechanical integrity and long-term durability [233]. The polymer membranes prepared by solvent casting/phase separation typically have tortuous pores of size >100 nm, medium distribution and high density [166]. Recently, Wang et al. have shown that fibro-porous coatings having large pore volumes, trans-membrane interconnected pore network, uniform and controllable porosity, pore sizes, thickness and hydrophilicity can be electrospun directly on the surface of miniature sensors [219, 220]. The polyurethane-based electrospun coatings were reported not only to function as mass-transport limiting membranes but also to have minimal effect on sensor sensitivity [220].

Polymer blends or multi-layered composites have also been used to obtain better controlled porosities, than monolithic polymers [166]. Uehara et al. proposed the use of nanofiltration composite membranes (with well controlled pore size in a range between 5 and 30 nm) made of polystyrene (PS) from PS-block-poly(methyl methacrylate) (PMMA) with homo-polymer PMMA to optimise the glucose diffusion and, at the same time, prevent large molecules (e.g., bovine serum albumin) to pass through [234]. To allow flexibility and, at the same time, limit the diffusion of glucose and hydrogen peroxide Chu et al. used porous polytetrafluoroethylene (PTFE) membrane with phospholipid polymer (2-methacryloyloxyethyl phosphorylcholine (MPC) copolymerized with 2-ethylhexylmethacrylate (EHMA) called PMEH which has a similar molecular configuration as a cell membrane [235]. Further, a novel, multifunctional membrane composed of MPC and n-butylmethacrylate (BMA) was proposed to improve the biocompatibility and sustain sensor performance [236].

Most polymers tested are porous. However, Updike's group reported the use of monolithic membranes based on polyurethane on glucose sensors. The monolithic membrane is non-porous, yet it acts as barrier due to its preferential dissolution and diffusion of solutes (rather than sieving the solutes through pores) [217]. The polyurethanes were based on block co-polymers having hydrophobic and hydrophilic domains (containing carboxylate groups and polyether segments) that limit solute transport. Another membrane that has hydrophobic and hydrophilic components is Nafion (a perfluorosulfonic acid polymer), whose sulfonic acid residues also repel anionic interferants [237, 238]. Moussy's group extensively tested both polyurethane and Nafion membranes as mass transport limiting membranes [170, 212, 227, 228, 232, 239–242]. In their studies, Moussy et al. reinforced polyurethane with epoxy resin for better mechanical and long-term durability. These membranes also have good mechanical properties, provide sufficient glucose limiting barrier and limit the amount of interferants passing through. Unfortunately, the properties of PU and Nafion based membranes are highly depended from the polymerisation/crosslinking conditions (e.g., temperature and humidity) [236, 243], which cause high within the production batch variability in sensor performance. Recently, Trzebinski et al. showed the efficacy of highly cross-linked neutral hydrogels based on HEMA, DHPMA and NVP crosslinked with EGDMA as mass-transport limiting membranes, whose performance is highly reproducible and not affected by the polymerization conditions [231].

Well organised, uniform and thin membrane made with high precision LBL assembly is another choice for mass transport limiting. Galeska et al. and more recently, Tipnis et al. used this technique to construct layers with humic acids/ferric cations (HAs/Fe3+), humic acids/poly diallyldimethyl-ammonium chloride (HAs/PDDA), and poly styrene sulfonate/poly dial-lyldimethylammonium chloride (PSS/PDDA) and optimised the diffusion of glucose and hydrogen peroxide through the membranes [244, 245]. Stein et al. assembled nanometre size (12 nm) layers of poly allylamine hydro-chloride (PAH) and poly sodium 4-styrenesulfonate (PSS) on glucose sen-sitive microspheres and by varying the assembly conditions and the number of layers adjusted the sensitivity to cover the hypo-, normo- and hyper-glycemic levels [70, 71]. However, the physical or ionic affinity binding in LBL assemblies can be susceptible to local variations in pH and ionic strength, often compromising the long-term stability of the membranes, especially when implanted.

Among optical systems based on affinity reactions, it is also important to entrap the affinity molecules (typically in the form of microspheres). To this goal, Ballerstadt et al. employed cellulose based microfiltration membrane with molecular mass cut off (MWCO) 6–8 k M_w which allowed the glucose and its oxidation products to freely pass through, while preventing the leak-age of the toxic Con A dye conjugate [69].

Modern fabrication methods such as lithography and ion-track etching for polymers; micromachining, anodization, powder sintering and sol-gel approaches for ceramics and semiconductor materials have been utilized to generate membranes with well controlled pore sizes ranging between 1 and 50 nm [166]. Such membranes, due to their molecular sieving abilities, func-tion as excellent mass-transport limiting structures. But, in practice, they are primarily used as enzyme or affinity molecule entrapping membranes than as mass-transport limiting membranes for glucose sensing [171–182].

Although most mass-transport limiting membranes aid sensors' functions for several months *ex-vivo* (without significant loss in performance), their *in vivo* functional life is severely limited by host tissue reactions. Despite long-standing research (Section 5.4), there still is a pressing need for tissue engineering coating strategies to reliably enhance the in vivo sensing lifetime of implantable biosensors.

5.4 Strategies to reliably extend the in vivo clinical life of implantable glucose biosensors

Foreign body reactions are detrimental to implanted sensors because they adversely affect the sensor's primary function of analyte sensing. As soon as the sensing device contacts biological fluids, a multitude of biomolecules, and then cells, foul its surface. The initial biofouling coat is usually 10 to 100 μm thick and it causes up to 50% reduction in analyte (glucose) permeability. The tissue inflammation reactions that follow biofouling further reduce glucose permeability by as much as 500%, as identified by Reichert's group, who suggest this needs to be countered more than biofouling [246]. On the other hand, Brauker et al. suggested that the barrier layer of cells formed by the compacting force of the fibrous tissue around the implant has a significant role in the decrease in sensor sensitivity and its eventual failure [247].

Considerable research has focused on reducing biofouling through the sensor's surface modification [248–250]. Regulatory agency approved biocompatible materials such as poly-tetrafluoroethylene (PTFE), silicone, polyester, polypropylene (PE), polyurethane (PU), polyethylene glycol (PEG), polyethylene oxide (PEO), polymethylmethacrylate (PMMA) and polyhydroxyethyl methacrylate (PHEMA), were used as off-the-shelf materials for the design of sensor coatings that have the dual roles of mass-transport limiting membrane and biocompatible surface. Such materials, although acceptable for most other prosthetic device applications, usually fail in preventing fibrous connective tissue build-up around the sensor. Thus, to minimize biofouling and tissue build-up, different surface modifications have been tested [217, 248, 249].

Surface functional group chemistry has significant role in biofouling. Typically, hydrophobic surfaces, such as $-CF_3$ and $-CH_3$ (with the exception of hexamethyl siloxyl groups) [251] and metal surfaces [252], such as gold, cause significantly higher protein adsorption and inflammatory cell adhesion compared to hydrophilic surfaces and sham controls [253]. Among the hydrophilic surfaces, $-COOH$ and $-OH$ groups induce lower protein adsorption and inflammatory cell infiltration compared to $-NH_2$ groups [251, 252]. Further, the fibrous capsule thickness for $-COOH$ and $-OH$ groups is usually similar to that of sham controls [252]. Among the hydrophilic polymers, Nafion is commonly tested for enzyme immobilization, interference elimination, mass-transport limitation, and biocompatibility [157, 212, 214, 215, 237, 241, 254]. But Nafion's sulfonate groups act as nuclei for calcium accumulation and calcium phosphate deposition (calcification) limiting their use as biocompatible surfaces for implants [240]. To avoid problems with ionic functional groups, hydrogels such as PHEMA, PEG and PVA that are neutral, polar, flexible and hydrophilic have been used to significantly lower biofouling, without calcification [249, 255–260]. Furthermore, biofouling was also lowered using biomimetic strategies of mimicking the natural membrane structure using lipids and lipid grafted polymers. These include phosphorylcholine, 2-methacryloyloxyethyl and polyurethane with phospholipid polar groups [235, 236, 261–263]. Hyaluronic acid, a natural hydrophilic (nonsulfated glycosaminoglycan) component of extracellular matrix, was also used to reduce biofouling [264].

Nano-textured surfaces have also been shown to improve biocompatibility through reducing biofouling. Nano-porous materials based on titania, silicon, anodic alumina and diamond-like carbon (DLC), due to their high aspect ratio features, are said to alter cell phenotype, proliferation and differentiation [250, 265–274]. Their use as biocompatible outer coatings on biosensors has been limited, but has potential.

Overall, the coatings based on hydrogels, phospholipids, hyaluronic acid, silica and other materials, having the correct surface chemistry and nano-structure that lower the degree of biofouling and inflammation, were shown to be non-toxic to surrounding tissues. Yet, they are susceptible to immune responses caused by the inevitable tissue injury during implantation that result in fibrous tissue build-up (fibrosis within weeks of implantation). The degree of fibrosis can further be intensified by mechanical injury due to implantation [275].

Recently, porous biomaterials were shown to not only disrupt fibrous capsule formation but also increase blood vascular supply [247, 276–280]. The materials tested as porous coatings for biosensors include poly(tetrafluoroethylene), silicone, polyester, polypropylene, polyurethane, poly(vinyl pyrrolidone), poly(vinyl alcohol), hydroxyethyl methacrylate, hydroxypropyl methacrylate, collagen-nordehydroglucuronic acid (NDGA) and poly L-lactic acid [77, 247, 255, 256, 277, 280–282]. Brauker et al. and Koshwanez et al. utilised synthetic polymers with low hydrophilicity as the porous backbone of coatings primarily aiming at improving blood vascular supply and disrupting fibrous capsule formation on immediate surface of the sensor [247, 277]. Other researchers have utilised porous hydrophilic polymers on their own [77, 255, 256, 280, 281]. However, the hydrophilicity deters cellular adhesion and transport into the coatings. The use of natural materials as a primary component of the coating is limited to the studies of Ju et al., who prepared semi-interpenetrating polymer networks of collagen interpenetrated with cross-linked NDGA [283, 284]. The semi-IPNs showed encouraging results but require additional testing to ascertain long-term *in vivo* functional efficacy [283, 284]. Most of the biocompatibility coatings tested were reported to increase the *in vivo* life of the biosensor but, often, only one of many supposedly identical sensors functions longer than a few weeks [277]. Hence, reliability in long-term performance of biosensors *in vivo* is still a concern. There is an urgent need for better coating materials and understanding the role of tissue-sensor coating material interactions on *in vivo* sensor failure.

Mimicking the fibro-porous architecture of natural extracellular matrix (ECM) is another approach that can be used to favourably modify the host's tissue responses to implanted biosensors. The efficacy of electrospun materials as ECM mimics has been extensively studied by Sanders et al. They showed that if the diameter of fibre is between 1 and 5.9 μm, most fibres did not cause fibrous capsule formation, and irrespective of surface charge or fibre diameters the degree of angiogenesis is not affected [285]. Electrospinning provides the versatility in controlling fibrous structures offering advantages such as high surface area to volume ratio, adjustable porosity, biomimic ECM structure and the ability to design nanofiber compositions having desired physicochemical and mechanical properties. But their application as tissue engineering (biocompatibility) coatings for implantable biosensors is limited. Only recently, we reported the use of electrospun PU-Gelatin co-axial fibre membranes as biomimetic coatings on miniature implantable glucose biosensors. Preliminary studies indicated better sensor sensitivity for PU-Ge co-axial fibre membrane coated sensors in the first four weeks compared to the traditional epoxy-PU coated base sensors [286].

6. Summary and future direction

The glucose sensing industry is primarily driven by the diabetes epidemic. The number of people diagnosed with diabetes in UK has more than doubled from 1.4 m in 1996 to 2.9 m in 2011, and is expected to cross 5 m by 2025 (Diabetes UK, 2013). Similarly, diabetes prevalence worldwide of 366 m in 2011 is expected to increase to 552 m by 2030 [287–289]. The long-standing goal for reliable continuous glucose monitoring is still unmet, the finger-prick home glucose monitoring devices are indicated as acceptable with an error margin of ±20%, and laboratory testing is still the gold standard for diabetes diagnosis. Nanomaterials have an important role in making glucose sensing technologies cross the frontiers of traditional electrochemical and optical methods through solving the problems with large error margins and reliable *in vivo* lifetime and function of implantable glucose biosensors. In addition, the need for glucose sensing is also rapidly increasing in food, cell culture and biotechnology industries that help sustain the growing human population.

Glucose is a generic sugar molecule whose detection in complex physiological media is often difficult because of the lack of selectivity and the competition from interferents. The electrochemical glucose detection methods also suffer due to poisoning of the electrode surface, wherein, ions and biological molecules bind to and block the electrode's catalytic sites for electrochemical reactions, while the optical methods often require unmasking of glucose-specific signal through complex data processing from background optical signals from the complex physiological media. Current commercial glucose sensing systems are strongly reliant on the use of glucose-specific binding molecules and enzymes for assays of blood glucose concentrations.

Molecular-scale glucose-specific interactions with binding molecules such as Con A, boronic acid derivatives, bacterial glucose binding proteins and enzymes are used as highly sensitive and specific optical tools for glucose detection. However, the binding molecules either do not show glucose specific optical activity or the signal is very weak. As a result they are conjugated with optically active molecules, e.g. organic fluorophores, for biosensor design. Conventionally, optical sensors measure glucose-induced changes in fluorescence emission or quenching signal (from the fluorophore label) due to change in conformation or competitive displacement, often using the highly sensitive FRET method involving dipole-dipole interactions.

To make fluorescence methods viable for *in vivo* CGM, nanomaterials and nano-fabrication methods are being employed [10, 290]. As summarised by Cash et al., [290] one approach uses a binding molecule and two fluorophores covalently bound on polymer nanospheres. When glucose binds to the binding molecule, the change in polymer conformation increases the average distance between the fluorophores, generating a FRET signal [291, 292]. In a second approach, hydrophobic binding molecules and a non-fluorescent dye that binds to the binding molecule to generate fluorescent signal are incorporated in the core of a hydrophobic polymer nanosphere. Glucose competitively displaces the non-fluorescent dye, thus reducing the fluorescence signal. The hydrophobic components inherently remain in the core of the nanospheres, thus making the sensors reversible [293]. A third approach uses layer-by-layer assembly of nanofilms on microspheres. Following the assembly, the microspheres are dissolved leaving microcapsules, in the lumen of which, the sensing components are loaded and encapsulated for glucose sensing [73, 294–296]. Irrespective of the sensing reactions, the nano- or micro-formulations, typically using near-infrared fluorophores, can be encapsulated within membranes and introduced to skin tissue as 'smart tattoos'. The tattoos are then irradiated with near-infrared light that penetrates the skin to sense glucose. This approach avoids the need for the frequent finger-pricking for conventional glucose monitoring. However, the disadvantage with the traditional fluorophores is that they photo-bleach and degrade causing gradual decrease in the measurable fluorescence signal.

An alternative for traditional fluorophores is the size-dependant change in optical properties for label-free detection of glucose using the nanomaterials: AuNPs, CNTs and QDs. They do not photo-bleach and when conjugated with glucose-specific binding molecules, they function as reversible sensors. AuNPs are known to exhibit size dependant optical absorption, which is exploited by covalent immobilization of glucose-binding agents on the surface of AuNPs [297, 298]. The AuNPs amplify the surface resonance Raman spectroscopy (SERS) signals and hence was tested for *in vivo* efficacy by implanting the surface-coated AuNPs in rats [33]. The near-infrared fluorescence of CNTs is used to measure glucose specific conformational changes in hydrogels incorporated with modified GOx and CNTs or aggregation of CNTs with Con A [299, 300]. The latter are encapsulated in microdialysis capillaries for in vivo glucose sensing [299]. QDs including CdTe, Mn-doped ZnS and TiO_2/SiO_2 are another group of nanomaterials that are conjugated with GOx, GDH, Con A or boronic acid derivatives for efficient glucose biosensing [82, 141, 144, 290, 301]. Thus the intrinsic optical properties of the nanomaterials: AuNPs, CNTs and especially QDs make them promising for the development of optical biosensors for long-term *in vivo* application e.g., as smart tattoos [290].

Inherently, electrochemical sensors are robust, economical, reliable, have wide detection limits, easy to miniaturise and require small sample volume. Nanomaterials further enhance their sensing efficiency and miniaturization. Broadly, the use of nanomaterials for improving working electrode design can be classified into three categories: one, the nanoporous electrode surface, two, the artificial mediators and three, the direct electrical wiring of the enzymes to electrode surface.

Nanoscale porosity (or roughness) at electrode surface enhances the sensor sensitivity through higher surface area available for catalysis and mass-transport. It also prevents the oxidation of interferents through a kinetically controlled electrochemical mechanism, and the electrode poisoning by chloride ions [11, 65, 66]. Nano-porosity on electrode surface is achieved either by etching or deposition of NPs. In the latter case, the NP composition can also be varied to lower the operating over-voltages for glucose sensing (Table 6-1) [40–44].

A major objective for the use of nanomaterials for glucose biosensing is to tune the operation over-voltages closer to 0.0 V. This not only reduces the energy (battery power) needed for sensor operation, but also prevents oxidation of common interfering agents such as ascorbic acid, uric acid and acetaminophen at the electrode surface. A wide range of artificial mediators are used to avoid the oxygen dependence of first generation glucose sensors, as well as to lower the operating over-voltages (Tables 4-1 & 6-1). The nano-sized mediators act as shuttles for electron transfer from the enzyme's redox centre to the electrode surface. The redox potentials of the mediators are usually close to that the redox centre of the enzyme, due to which the overall sensor operating over-voltages are low [28]. To further increase the efficiency of capturing the electrons from the reduced enzyme, the artificial mediators are covalently coupled with the enzyme to make electron relays that the enzyme communicates directly with the electrode [22]. However, due to their nano-size, it is difficult to contain the mediators or electron relays at the electrode surface, and if implanted in the body, the free diffusing redox mediators are highly toxic.

For direct electrical wiring of enzyme's redox centre to the electrode surface, the enzyme-mediator electron relays are further covalently attached to redox polymers, which in turn are attached to electrode surface [1, 23, 160]. Such covalent immobilization not only solves the problem of freely diffusing redox mediators, but also makes the electron transfer kinetics between enzyme and electrode highly efficient. Multiple layers of enzymes can be attached to tiny electrode surface to generate huge current densities. Added with the low operational over-voltage (0.0 to 0.1 V) and negligible interference, electrically wired systems are attractive for in vivo glucose sensing.

Another key component for sensor function is the semi-permeable membrane (also known as flux-limiting or mass-transport-limiting membrane), which determines the linear-detection range for the sensor. Off-the-shelf nanoporous dialysis membranes (e.g., cellulose acetate) and custom designed polymers (e.g., polyurethane and Nafion) are commonly used as semi-permeable membranes. Lithography techniques can also be used to develop nanoporous membranes for biosensors [166–168]. Ancillary functions for this membrane include eliminating interference and entrapping or immobilizing of glucose-specific affinity molecules. Layer-by-layer self-assembly and electro-polymerisation, using polyelectrolytes and conducting polymers is used for immobilizing multiple components, and for precision controlled tailoring of membrane thickness and sensor sensitivity. Metal oxides and NPs are also conjugated to these membranes for preventing interference [171–187].

Table 6-1 Examples for tuning the over-voltages for electrochemical sensors using nanomaterials. For direct oxidation of glucose, nanomaterials typically lower the oxidation potential. However, for H_2O_2 detection, often the same nanomaterial can be used to both oxidize and reduce H_2O_2. Where possible, the reduction potential (−0.2 to 0.0 V) is chosen, because this practically eliminates interference from common redox interferents such as ascorbic acid, uric acid and acetaminophen

Nano-material	Enzyme	Over-voltage (V) Oxidation	Reduction	Ref
Direct oxidation of glucose:				
Ag-Au alloy, Ag & Au NPs	–	0.3	–	[40, 302]
Cu microparticles on Au electrode	–	0.35	–	[41]
Ag$_2$O nanowalls on Cu substrate	–	0.4	–	[42]
Porous CuO films on Pt/Ti/Si substrates	–	0.65	–	[43]
Ni NPs on graphite-like carbon	–	0.4	–	[46]
MWNTs	–	0.2	–	[47]
Pt-Ir alloy NPs	–	0.1	–	[303]
Cu NPs + CNT	–	0.65	–	[50–52]
Pt NPs + MWNTs	–	0.0	–	[54]
BO NPs + MWNTs	–	0.122	–	[55]
Pd NPs + Functionalised CNTs	–	0.4	–	[58]
Oxidation of H_2O_2:				
Ir Oxide	GOx	0.4	−0.1	[102–104]
Polypyrrole	GOx, HRP	0.15	–	[105, 106]
Ferrocene	GOx, HRP	0.0	–	[107]
Purssian Blue	GOx	–	−0.1	[120, 121]
Ru, Rh, Ir, Pt-Ru + carbon paste	GOx	0.0	−0.1	[99, 100, 123]
Pt, Pd NPs + graphite	GOx	–	−0.05	[124]
Fe(CN)$_6$ + CNTs + Chitosan	GOx	–	−0.2	[125]
Pt NPs, CNTs + Nafion	GOx	–	−0.05	[127]
Au NPs + Silicate	GOx	0.3	−0.2	[128]
Cu^{2+} + DNA-Polyallyamine polyion complex	GOx	–	−0.2	[129]

Specialized membranes are also used for enhancing the biocompatibility of the glucose biosensors for implantation. Surface modification of the semi-permeable membrane has been used for enhancing biocompatibility. Often this doesn't prevent fibrous encapsulation of the device. Additional outer membrane based on porous polymers and electrospun membranes help prevent the fibrous tissue build-up on the immediate sensor surface [219, 220, 247, 276–280, 283, 284]. Further, biomimetic fibroporous-coatings and sensor miniaturization could play an important role in addressing the problem of sensor failure when implanted [219, 220, 286].

Overall, nanomaterials help address the problems, for conventional optical and electrochemical biosensors, by enhancing the preferential detection of glucose or its oxidation products through better electron transfer kinetics, sensitivity and response time, while lowering the operating over-voltages for energy efficiency and avoid interference. The reproducible production of nano-materials and nano-structures at low cost is vital for the successful development of nano-technologies for glucose sensing. Several products, especially, home glucose monitoring devices, use nano-materials, but the need for reliable long-term CGM is still unmet. Nano-materials and nano-technologies have an important role in achieving the long-awaited CGM technology.

7. References

1. Heller, A., Feldman, B., 2008, "Electrochemical glucose sensors and their applications in diabetes management," *Chemical Reviews*, **108**, pp. 2482–2505.
2. Tonyushkina, K., Nichols, J. H., 2009, "Glucose meters a review of technical challenges to obtaining accurate results," *Journal of Diabetes Science and Technology*, **3**, pp. 971–980.
3. Kondepati, V. R., Heise, H. M., 2007, "Recent progress in analytical instrumentation for glycemic control in diabetic and critically ill patients," *Analytical and Bioanalytical Chemistry*, **388**, pp. 545–563.
4. Oliver, N. S., Toumazou, C., Cass, A. E., Johnston, D. G., 2009, "Glucose sensors a review of current and emerging technology," *Diabetes Medicine*, **26**, pp. 197–210.
5. Queinnec, I., Destruhaut, C., Pourciel, J. B., Goma, G., 1992, "An effective automated glucose sensor for fermentation monitoring and control," *World Journal of Microbiology and Biotechnology*, **8**, pp. 7–13.
6. Brooks, S. L., Ashby, R. E., Turner, A. P. F., Calder, M. R., Clarke, D. J., 1987, "Development of an on-line glucose sensor for fermentation monitoring," *Biosensors*, **3**, pp. 45–56.
7. Tothill, I. E., Newman, J. D., White, S. F., Turner, A. P. F., 1997, "Monitoring of the glucose concentration during microbial fermentation using a novel mass-producible biosensor suitable for on-line use," *Enzyme and Microbial Technology*, **20**, pp. 590–596.
8. Lee, M.-C., Kabilan, S., Hussain, A., Yang, X., Blyth, J., Lowe, C. R., 2004, "Glucose-sensitive holographic sensors for monitoring bacterial growth," *Analytical Chemistry*, **76**, pp. 5748–5755.
9. Kimura, H., Takeyama, H., Komori, K., Yamamoto, T., Sakai, Y., Fujii, T., 2010, "Microfluidic device with integrated glucose sensor for cell-based assay in toxicology," *Journal of Robotics and Mechatronics*, **22**, pp. 594–600.
10. Pickup, J. C., Hussain, F., Evans, N. D., Rolinski, O. J., Birch, D. J., 2005, "Fluorescence-based glucose sensors," *Biosensors Bioelectronics*, **20**, pp. 2555–2565.
11. Park, S., Boo, H., Chung, T. D., 2006, "Electrochemical non-enzymatic glucose sensors," *Analytica Chimica Acta*, **556**, pp. 46–57.
12. Ernst, S., Hamann, C. H., Heitbaum, J., 1980, "Electrooxidation of glucose in phosphate buffer solutions kinetics and reaction mechanism," *Berichte der Bunsengesellschaft/Physical Chemistry Chemical Physics*, **84**, pp. 50–55.
13. Beden, B., Largeaud, F., Kokoh, K. B., Lamy, C., 1996, "Fourier transform infrared reflectance spectroscopic investigation of the electrocatalytic oxidation of glucose identification of reactive intermediates and reaction products," *Electrochimica Acta*, **41**, pp. 701–709.
14. Ernst, S., Heitbaum, J., Hamann, C. H., 1979, "The electrooxidation of glucose in phosphate buffer solutions part I. Reactivity and kinetics below 350 mV/RHE," *Journal of Electroanalytical Chemistry*, **100**, pp. 173–183.
15. Ivnitski, D., Branch, B., Atanassov, P., Apblett, C., 2006, "Glucose oxidase anode for biofuel cell based on direct electron transfer," *Electrochemical Communication*, **8**, pp. 1204–1210.

16. Hecht, H. J., Kalisz, H. M., Hendle, J., Schmid, R. D., Schomburg, D., 1993, "Crystal-structure of glucose-oxidase from aspergillus-niger refined at 2.3 angstrom resolution," *Journal of Molecular Biology*, **229**, pp. 153–172.

17. Wang, J., 2008, "Electrochemical glucose biosensors," *Chemical Reviews* **108**, pp. 814–825.

18. Henning, T., 2010, "Commercially available continous glucose monitoring systems," *In vivo glucose sensing*, Cunningham, D. D., Stenken, J. A., eds, John Wiley & Sons, Inc., Hoboken, New Jersey, pp. 113–156.

19. Clark, L. C., 1959, "Electrochemical device for chemical analysis," US Patent No. 2913386.

20. Gough, D. A., Kumosa, L. S., Routh, T. L., Lin, J. T., Lucisano, J. Y., 2010, "Function of an implanted tissue glucose sensor for more than 1 year in animals," *Science Translational Medicine*, **2**, pp. 42–53.

21. U. S. Food and Drugs Administration, 2009, "GDH-PQQ (glucose dehydrogenase pyrroloquinoline quinone) Glucose Monitoring Technology," http//www.fda.gov/Safety/MedWatch/SafetyInformation/SafetyAlertsforHumanMedicalProducts/ucm177295.htm.

22. Heller, A., 1990, "Electrical wiring of redox enzymes," *Accounts of Chemical Research*, **23**, pp. 128–134.

23. Heller, A., 1992, "Electrical connection of enzyme redox centers to electrodes," *The Journal of Physical Chemistry*, **96**, pp. 3579–3587.

24. Degani, Y., Heller, A., 1988, "Direct electrical communication between chemically modified enzymes and metal electrodes. 2. Methods for bonding electron-transfer relays to glucose oxidase and D-amino-acid oxidase," *Journal of American Chemical Society*, **110**, pp. 2615–2620.

25. Degani, Y., Heller, A., 1987, "Direct electrical communication between chemically modified enzymes and metal electrodes. I. Electron transfer from glucose oxidase to metal electrodes via electron relays, bound covalently to the enzyme," *The Journal of Physical Chemistry*, **91**, pp. 1285–1289.

26. Yehezkeli, O., Yan, Y. M., Baravik, I., Tel-Vered, R., Willner, I., 2009, "Integrated oligoaniline-cross-linked composites of Au nanoparticles/glucose oxidase electrodes a generic paradigm for electrically contacted enzyme systems," *Chemistry A European Journal*, **15**, pp. 2674–2679.

27. Yehezkeli, O., Ovits, O., Tel-Vered, R., Raichlin, S., Willner, I., 2010, "Reconstituted enzymes on electropolymerizable FAD-modified metallic nanoparticles functional units for the assembly of effectively "wired" enzyme electrodes," *Electroanalysis*, **22**, pp. 1817–1823.

28. Chaubey, A., Malhotra, B. D., 2002, "Mediated biosensors," *Biosensors Bioelectronics*, **17**, pp. 441–456.

29. Shen, Y. C., Davies, A. G., Linfield, E. H., Elsey, T. S., Taday, P. F., Arnone, D. D., 2003, "The use of Fourier-transform infrared spectroscopy for the quantitative determination of glucose concentration in whole blood," *Physics in Medicine and Biology*, **48**, pp. 2023–2032.

30. Nelson, L. A., McCann, J. C., Loepke, A. W., Wu, J., Ben Dor, B., Kurth, C. D., 2006, "Development and validation of a multiwavelength spatial

domain near-infrared oximeter to detect cerebral hypoxia-ischemia," *Journal of Biomedical Optics*, **11**, pp. 064022.

31. Cameron, B. D., Anumula, H., 2006, "Development of a real-time corneal birefringence compensated glucose sensing polarimeter," *Diabetes Technology and Therapeutics*, **8**, pp. 156–164.

32. Esenaliev, R. O., Larin, K. V., Larina, I. V., Motamedi, M., 2001, "Noninvasive monitoring of glucose concentration with optical coherence tomography," *Optics Letters*, **26**, pp. 992–994.

33. Dieringer, J. A., McFarland, A. D., Shah, N. C., Stuart, D. A., Whitney, A. V., Yonzon, C. R., Young, M. A., Zhang, X., Van Duyne, R. P., 2006, "Surface enhanced Raman spectroscopy new materials, concepts, characterization tools, and applications," *Faraday Discuss*, **132**, pp. 9–26.

34. Ashok, A., Nirmalkumar, A., Jeyashanthi, N., 2010, "A novel method for blood glucose measurement by noninvasive technique using laser," *International Journal of Biological and Medical Sciences*, **5**, pp. 58–64.

35. Clarke, W. L., Cox, D., Gonder-Frederick, L. A., Carter, W., Pohl, S. L., 1987, "Evaluating clinical accuracy of systems for self-monitoring of blood glucose," *Diabetes Care*, **10**, pp. 622–628.

36. Shah, N. C., Yuen, J. M., Lyandres, O., Glucksberg, M. R., Walsh, J. T., Van Duyne, R. P., 2010, "Surface-enhanced Raman spectroscopy for glucose sensing," *In vivo glucose sensing*, Cunningham, D. D., Stenken, J. A., eds, John Wiley & Sons, Inc., Hoboken, New Jersey, pp. 421–443.

37. Weiss, R., Yegorchikov, Y., Shusterman, A., Raz, I., 2007, "Noninvasive continuous glucose monitoring using photoacoustic technology-results from the first 62 subjects," *Diabetes Technology and Therapeutics*, **9**, pp. 68–74.

38. Pfutzner, A., Caduff, A., Larbig, M., Schrepfer, T., Forst, T., 2004, "Impact of posture and fixation technique on impedance spectroscopy used for continuous and noninvasive glucose monitoring," *Diabetes Technology and Therapeutics*, **6**, pp. 435–441.

39. Sun, Y., Buck, H., Mallouk, T. E., 2001, "Combinatorial discovery of alloy electrocatalysts for amperometric glucose sensors," *Analytical Chemistry*, **73**, 1599–1604.

40. Tominaga, M., Shimazoe, T., Nagashima, M., Kusuda, H., Kubo, A., Kuwahara, Y., Taniguchi, I., 2006, "Electrocatalytic oxidation of glucose at gold-silver alloy, silver and gold nanoparticles in an alkaline solution," *Journal of Electroanalytical Chemistry*, **590**, pp. 37–46.

41. Casella, I. G., Gatta, M., Guascito, M. R., Cataldi, T. R. I., 1997, "Highly-dispersed copper microparticles on the active gold substrate as an amperometric sensor for glucose," *Analytica Chimica Acta*, **357**, pp. 63–71.

42. Fang, B., Gu, A., Wang, G., Wang, W., Feng, Y., Zhang, C., Zhang, X., 2009, "Silver oxide nanowalls grown on Cu substrate as an enzymeless glucose sensor," *ACS Applied Materials and Interfaces*, **1**, pp. 2829–2834.

43. Cherevko, S., Chung, C. H., 2010, "The porous CuO electrode fabricated by hydrogen bubble evolution and its application to highly sensitive non-enzymatic glucose detection," *Talanta*, **80**, pp. 1371–1377.

44. Xu, Q., Zhao, Y., Xu, J. Z., Zhu, J.-J., 2006, "Preparation of functionalized copper nanoparticles and fabrication of a glucose sensor," *Sensors and Actuators B – Chemistry*, **114**, pp. 379–386.

45. Sakamoto, M., Takamura, K., 1982, "Catalytic oxidation of biological components on platinum electrodes modified by adsorbed metals anodic oxidation of glucose," *Bioelectrochemistryistry Bioenergetics*, **9**, pp. 571–582.

46. You, T., Niwa, O., Chen, Z., Hayashi, K., Tomita, M., Hirono, S., 2003, "An amperometric detector formed of highly dispersed Ni nanoparticles embedded in a graphite-like carbon film electrode for sugar determination," *Analytical Chemistry*, **75**, pp. 5191–5196.

47. Ye, J.-S., Wen, Y., De Zhang, W., Ming Gan, L., Xu, G. Q., Sheu, F.-S., 2004, "Nonenzymatic glucose detection using multi-walled carbon nanotube electrodes," *Electrochemical Communications*, **6**, pp. 66–70.

48. Jianxiong, W., Xiaowei, S., Xianpeng, C., Yu, L., Li, S., ShiSheng, X., 2007, "Nonenzymatic glucose sensor using freestanding single-wall carbon nanotube films," *Electrochemical and Solid State Letters*, **10**, pp. J58–J60.

49. Tan, C. K., Loh, K. P., John, T. T. L., 2008, "Direct amperometric detection of glucose on a multiple-branching carbon nanotube forest," *Analyst*, **133**, pp. 448–451.

50. Yang, J., Jiang, L. C., Zhang, W. D., Gunasekaran, S., 2010, "A highly sensitive non-enzymatic glucose sensor based on a simple two-step electrodeposition of cupric oxide (CuO) nanoparticles onto multi-walled carbon nanotube arrays," *Talanta*, **82**, pp. 25–33.

51. Jiang, L. C., Zhang, W. D., 2010, "A highly sensitive nonenzymatic glucose sensor based on CuO nanoparticles-modified carbon nanotube electrode," *Biosensors Bioelectronics*, **25**, pp. 1402–1407.

52. Kang, X., Mai, Z., Zou, X., Cai, P., Mo, J., 2007, "A sensitive nonenzymatic glucose sensor in alkaline media with a copper nanocluster/multiwall carbon nanotube-modified glassy carbon electrode," *Analytical Biochemistry*, **363**, pp. 143–150.

53. Zhu, H., Lu, X., Li, M., Shao, Y., Zhu, Z., 2009, "Nonenzymatic glucose voltammetric sensor based on gold nanoparticles/carbon nanotubes/ionic liquid nanocomposite," *Talanta*, **79**, pp. 1446–1453.

54. Rong, L.-Q., Yang, C., Qian, Q.-Y., Xia, X.-H., 2007, "Study of the non-enzymatic glucose sensor based on highly dispersed Pt nanoparticles supported on carbon nanotubes," *Talanta*, **72**, pp. 819–824.

55. Wang, Y., Zhang, D., Zhang, W., Gao, F., Wang, L., 2009, "A facile strategy for nonenzymatic glucose detection," *Analytical Biochemistry*, **385**, pp. 184–186.

56. Chen, J., Zhang, W.-D., Ye, J.-S., 2008, "Nonenzymatic electrochemical glucose sensor based on MnO$_2$/MWNTs nanocomposite," *Electrochemical Communications*, **10**, pp. 1268–1271.

57. Cui, H.-F., Ye, J.-S., Liu, X., Zhang, W.-D., Sheu, F.-S., 2006, "Pt–Pb alloy nanoparticle/carbon nanotube nanocomposite a strong electrocatalyst for glucose oxidation," *Nanotechnology*, **17**, pp. 2334–2339.

58. Chen, X. M., Lin, Z. J., Chen, D. J., Jia, T. T., Cai, Z. M., Wang, X. R., Chen, X., Chen, G. N., Oyama, M., 2010, "Nonenzymatic amperometric sensing

of glucose by using palladium nanoparticles supported on functional carbon nanotubes," *Biosensors Bioelectronics*, **25**, pp. 1803–1808.

59. Liu, D., Luo, Q., Zhou, F., 2010, "Nonenzymatic glucose sensor based on gold-copper alloy nanoparticles on defect sites of carbon nanotubes by spontaneous reduction," *Synthetic Metals*, **160**, pp. 1745–1748.

60. Zhang, Y., Xiao, X., Sun, Y., Shi, Y., Dai, H., Ni, P., Hu, J., Li, Z., Song, Y., Wang, L., 2013, "Electrochemical deposition of nickel nanoparticles on reduced graphene oxide film for nonenzymatic glucose sensing," *Electroanalysis*, **25**, pp. 959–966.

61. Gao, H., Xiao, F., Ching, C. B., Duan, H., 2011, "One-step electrochemical synthesis of PtNi nanoparticle-graphene nanocomposites for nonenzymatic amperometric glucose detection," *ACS Applied Materials & Interfaces*, **3**, pp. 3049–3057.

62. Zhang, Y., Wang, Y., Jia, J., Wang, J., 2012, "Nonenzymatic glucose sensor based on graphene oxide and electrospun NiO nanofibers," *Sensors and Actuators B Chemical*, **171–172**, pp. 580–587.

63. Zhang, Y., Xu, F., Sun, Y., Shi, Y., Wen, Z., Li, Z., 2011, "Assembly of $Ni(OH)_2$ nanoplates on reduced graphene oxide a two dimensional nanocomposite for enzyme-free glucose sensing," *Journal of Materials Chemistry*, **21**, pp. 16949–16957.

64. Dong, X.-C., Xu, H., Wang, X.-W., Huang, Y.-X., Chan-Park, M. B., Zhang, H., Wang, L.-H., Huang, W., Chen, P., 2012, "3D graphene-cobalt oxide electrode for high-performance supercapacitor and enzymeless glucose detection," *ACS Nano*, **6**, pp. 3206–3213.

65. Park, S., Chung, T. D., Kim, H. C., 2003, "Nonenzymatic glucose detection using mesoporous platinum," *Analytical Chemistry*, **75**, pp. 3046–3049.

66. Boo, H., Park, S., Ku, B., Kim, Y., Park, J. H., Kim, H. C., Chung, T. D., 2004, "Ionic strength-controlled virtual area of mesoporous platinum electrode," *Journal of American Chemical Society*, **126**, pp. 4524–4525.

67. Shoji, E., Freund, M. S., 2001, "Potentiometric sensors based on the inductive effect on the pK(a) of poly(aniline) a nonenzymatic glucose sensor," *Journal of American Chemical Society*, **123**, pp. 3383–3384.

68. Joo, S., Park, S., Chung, T. D., Kim, H. C., 2007, "Integration of a nanoporous platinum thin film into a microfluidic system for non-enzymatic electrochemical glucose sensing," *Analytical Science*, **23**, pp. 277–281.

69. Ballerstadt, R., Polak, A., Beuhler, A., Frye, J., 2004, "In vitro long-term performance study of a near-infrared fluorescence affinity sensor for glucose monitoring," *Biosensors Bioelectronics*, **19**, pp. 905–914.

70. Stein, E. W., Grant, P. S., Zhu, H., McShane, M. J., 2007, "Microscale enzymatic optical biosensors using mass transport limiting nanofilms. 1. Fabrication and characterization using glucose as a model analyte," *Analytical Chemistry*, **79**, pp. 1339–1348.

71. Stein, E. W., Singh, S., McShane, M. J., 2008, "Microscale enzymatic optical biosensors uing mass transport limiting nanofilms. 2. Response modulation by varying analyte transport properties," *Analytical Chemistry*, **80**, pp. 1408–1417.

72. Chaudhary, A., Harma, H., Hanninen, P., McShane, M. J., Srivastava, R., 2011, "Glucose response of near-infrared alginate-based microsphere sensors under dynamic reversible conditions," *Diabetes Technology and Therapeutics*, **13**, pp. 827–835.

73. Chaudhary, A., Raina, M., Harma, H., Hanninen, P., McShane, M. J., Srivastava, R., 2009, "Evaluation of glucose sensitive affinity binding assay entrapped in fluorescent dissolved-core alginate microspheres," *Biotechnology Bioengineering*, **104**, pp. 1075–1085.

74. Chaudhary, A., Raina, M., McShane, M. J., Srivastava, R., 2009, "Dissolved core alginate microspheres as "smart-tattoo" glucose sensors," *Conference Proceedings of IEEE Engineering in Medicine and Biology Society*, **2009**, pp. 4098–4101.

75. Srivastava, R., Jayant, R. D., Chaudhary, A., McShane, M. J., 2011, ""Smart tattoo" glucose biosensors and effect of coencapsulated anti-inflammatory agents," *Journal of Diabetes Science and Technology*, **5**, pp. 76–85.

76. Meadows, D. L., Schultz, J. S., 1993, "Design, manufacture and characterization of an optical fiber glucose affinity sensor based on an homogeneous fluorescence energy transfer assay system," *Analytica Chimica Acta*, **280**, pp. 21–30.

77. Marshall, A., Irvin, C., Barker, T., Sage, E., Hauch, K., Ratner B., 2004, "Biomaterials with tightly controlled pore size that promote vascular ingrowth," *Abstract of Papers American Chemical Society*, **45**, pp. 100–101.

78. Alexeev, V. L., Das, S., Finegold, D. N., Asher, S. A., 2004, "Photonic crystal glucose-sensing material for noninvasive monitoring of glucose in tear fluid," *Clinical Chemistry*, **50**, pp. 2353–2360.

79. Kabilan, S., Marshall, A. J., Sartain, F. K., Lee, M.-C., Hussain, A., Yang, X., Blyth, J., Karangu, N., James, K., Zeng, J., Smith, D., Domschke, A., Lowe, C. R., 2005, "Holographic glucose sensors," *Biosensors Bioelectronics*, **20**, pp. 1602–1610.

80. Pickup, J. C., Zhi, Z. L., Khan, F., Saxl, T., Birch, D. J., 2008, "Nanomedicine and its potential in diabetes research and practice," *Diabetes Metabolism Research and Reviews*, **24**, pp. 604–610.

81. Cordes, D. B., Gamsey, S., Singaram, B., 2006, "Fluorescent quantum dots with boronic acid substituted viologens to sense glucose in aqueous solution," *Angewandte Chemie International Edition*, **45**, pp. 3829–2832.

82. Tang, B., Cao, L., Xu, K., Zhuo, L., Ge, J., Li, Q., Yu, L., 2008, "A new nanobiosensor for glucose with high sensitivity and selectivity in serum based on fluorescence resonance energy transfer (FRET) between CdTe quantum dots and Au nanoparticles," *Chemistry*, **14**, pp. 3637–3644.

83. Clark, L. C. Jr., Lyons, C., 1962, "Electrode systems for continuous monitoring in cardiovascular surgery," *Annals of New York Academy of Science*, **102**, pp. 29–45.

84. Suzuki, H., Tamiya, E., Karube, I., 1988, "Fabrication of an oxygen-electrode using semiconductor technology," *Analytical Chemistry*, **60**, pp. 1078–1080.

85. Suzuki, H., Sugama, A., Kojima, N., Takei, F., Ikegami, K., 1991, "A miniature clark-type oxygen-electrode using a polyelectrolyte and its application as a glucose sensor," *Biosensors Bioelectronics*, **6**, pp. 395–400.

86. Suzuki, H., Arakawa, H., Karube, I., 2001, "Fabrication of a sensing module using micromachined biosensors," *Biosensors Bioelectronics,* **16,** pp. 725–733.

87. Mitsubayashi, K., Wakabayashi, Y., Tanimoto, S., Murotomi, D., Endo, T., 2003, "Optical-transparent and flexible glucose sensor with ITO electrode," *Biosensors Bioelectronics,* **19,** pp. 67–71.

88. Wang, J., Lu, F., 1998, "Oxygen-rich oxidase enzyme electrodes for operation in oxygen-free solutions," *Journal of American Chemical Society,* **120,** pp. 1048–1050.

89. Wang, J., Mo, J. W., Li, S. F., Porter, J., 2001, "Comparison of oxygren-rich and mediator-based glucose-oxidase carbon-paste electrodes," *Analytica Chimica Acta,* **441,** pp. 183–189.

90. Gough, D. A., Lucisano, J. Y., Tse, P. H. S., 1985, "Two-dimensional enzyme electrode sensor for glucose," *Analytical Chemistry,* **57,** pp. 2351–2357.

91. Reach, G., Wilson, G. S., 1992, "Can continuous glucose monitoring be used for the treatment of diabetes," *Analytical Chemistry,* **64,** pp. A381–A386.

92. Armour, J. C., Lucisano, J. Y., McKean, B. D., Gough, D. A., 1990, "Application of chronic intravascular blood-glucose sensor in dogs," *Diabetes,* **39,** pp. 1519–1526.

93. Wang, J., Li, S., Mo, J.-W., Porter, J., Musameh, M. M., Dasgupta, P. K., 2002, "Oxygen-independent poly(dimethylsiloxane)-based carbon-paste glucose biosensors," *Biosensors Bioelectronics,* **17,** pp. 999–1003.

94. Schaffar, B. P. H., Wolfbeif, O. S., 1990, "A fast responding fibre optic glucose biosensor based on an oxygen optrode," *Biosensors Bioelectronics,* **5,** pp. 137–148.

95. Moreno-Bondi, M. C., Wolfbeis, O. S., Leiner, M. J. P., Schaffar, B. P. H., 1990, "Oxygen optrode for use in a fiber-optic glucose biosensor," *Analytical Chemistry,* **62,** pp. 2377–2380.

96. Rosenzweig, Z., Kopelman, R., 1996, "Analytical properties and sensor size effects of a micrometer-sized optical fiber glucose biosensor," *Analytical Chemistry,* **68,** pp. 1408–1413.

97. Li, L., Walt, D. R., 1995, "Dual-analyte fiber-optic sensor for the simultaneous and continuous measurement of glucose and oxygen," *Analytical Chemistry,* **67,** pp. 3746–3752.

98. Jacobs, C. B., Peairs, M. J., Venton, B. J., 2010, "Review carbon nanotube based electrochemical sensors for biomolecules," *Analytica Chimica Acta,* **662,** pp. 105–127.

99. Wang, J., Liu, J., Chen, L., Lu, F., 1994, "Highly selective membrane-free, mediator-free glucose biosensor," *Analytical Chemistry,* **66,** pp. 3600–3603.

100. Liu, J., Lu, F., Wang, J., 1999, "Metal-alloy-dispersed carbon-paste enzyme electrodes for amperometric biosensing of glucose," *Electrochemical Communications,* **1,** pp. 341–344.

101. Wang, J. J., Myung, N. V., Yun, M. H., Monbouquette, H. G., 2005, "Glucose oxidase entrapped in polypyrrole on high-surface-area pt electrodes a model platform for sensitive electroenzymatic biosensors," *Journal of Electroanalytical Chemistry,* **575,** pp. 139–146.

102. Elzanowska, H., Abu-Irhayem, E., Skrzynecka, B., Birss, V. I., 2004, "Hydrogen peroxide detection at electrochemically and sol-gel derived Ir oxide films," *Electroanalysis*, **16**, pp. 478–490.
103. Cox, J. A., Lewinski, K., 1993, "Flow-injection amperometric determination of hydrogen-peroxide by oxidation at an iridium oxide electrode," *Talanta*, **40**, pp. 1911–1915.
104. Rauh, R. D., Twardoch, U. M., Jones, G. S., Jones, R. B., 1999, "Iridium oxide vertical bar glucose oxidase biosensors and microprobes," Oyama, N., Birss, V., eds, *Proceedings of the Symposium on Molecular Functions of Electroactive Thin Films*, vol. **98**, Electrochemical Society Inc, Pennington, pp. 176–183.
105. Tatsuma, T., Watanabe, T., Watanabe, T., 1993, "Electrochemical characterization of polypyrrole bienzyme electrodes with glucose-oxidase and peroxidase," *Journal of Electroanalytical Chemistry*, **356**, pp. 245–253.
106. DeBenedetto, G. E., Palmisano, F., Zambonin, P. G., 1996, "One-step fabrication of a bienzyme glucose sensor based on glucose oxidase and peroxidase immobilized onto a poly(pyrrole) modified glassy carbon electrode," *Biosensors Bioelectronics*, **11**, pp. 1001–1008.
107. Ohara, T. J., Vreeke, M. S., Battaglini, F., Heller, A., 1993, "Bienzyme sensors based on electrically wired peroxidase," *Electroanalysis*, **5**, pp. 825–831.
108. Tian, F. M., Zhu, G. Y., 2002, "Bienzymatic amperometric biosensor for glucose based on polypyrrole/ceramic carbon as electrode material," *Analytica Chimica Acta*, **451**, pp. 251–258.
109. Elekes, O., Moscone, D., Venema, K., Korf, J., 1995, "Bienzyme reactor for electrochemical detection of low concentrations of uric-acid and glucose," *Clinica Chimica Acta*, **239**, pp. 153–165.
110. Mulchandani, A., Pan, S. T., 1999, "Ferrocene-conjugated m-phenylenediamine conducting polymer-incorporated peroxidase biosensors," *Analytical Biochemistry*, **267**, pp. 141–147.
111. Liu, H. Y., Ying, T. L., Sun, K., Li, H. H., Qi, D. Y., 1997, "Reagentless amperometric biosensors highly sensitive to hydrogen peroxide, glucose and lactose based on N-methyl phenazine methosulfate incorporated in a Nafion film as an electron transfer mediator between horseradish peroxidase and an electrode," *Analytica Chimica Acta*, **344**, pp. 187–199.
112. Shan, D., Cosnier, S., Mousty, C., 2003, "HRP wiring by redox active layered double hydroxides application to the mediated H_2O_2 detection," *Analytical Letters*, **36**, pp. 909–922.
113. Asberg, P., Inganas, O., 2003, "Hydrogels of a conducting conjugated polymer as 3-D enzyme electrode," *Biosensors Bioelectronics*, **19**, pp. 199–207.
114. Gamburzev, S., Atanasov, P., Ghindilis, A. L., Wilkins, E., Kaisheva, A., Iliev, I., 1997, "Bifunctional hydrogen peroxide electrode as an amperometric transducer for biosensors," *Sensor and Actuators B – Chemistry*, **43**, pp. 70–77.
115. Gao, Z. Q., Binyamin, G., Kim, H. H., Barton, S. C., Zhang, Y. C., Heller, A., 2002, "Electrodeposition of redox polymers and co-electrodeposition of enzymes by coordinative crosslinking," *Angewandte Chemie International Edition*, **41**, pp. 810–813.
116. Kenausis, G., Chen, Q., Heller, A., 1997, "Electrochemical glucose and lactate sensors based on "wired" thermostable soybean peroxidase operating

continuously and stably at 37 degrees C," *Analytical Chemistry*, **69**, pp. 1054–1060.

117. Calvo, E. J., Battaglini, F., Danilowicz, C., Wolosiuk, A., Otero, M., 2000, "Layer-by-layer electrostatic deposition of biomolecules on surfaces for molecular recognition, redox mediation and signal generation," *Faraday Discuss*, **116**, pp. 47–65.

118. Solna, R., Dock, E., Christenson, A., Winther-Nielsen, M., Carlsson, C., Emneus, J., Ruzgas, T., Skladal, P., 2005, "Amperometric screen-printed biosensor arrays with co-immobilised oxidoreductases and cholinesterases," *Analytica Chimica Acta*, **528**, pp. 9–19.

119. Karyakin, A. A., 2001, "Prussian blue and its analogue electrochemistry and analytical applications," *Electroanalysis*, **13**, pp. 813–819.

120. Karyakin, A. A., Karyakina, E. E., Gorton, L., 1998, "The electrocatalytic activity of Prussian blue in hydrogen peroxide reduction studied using a wall-jet electrode with continuous flow," *J Electroanalysisytical Chemistry*, **456**, pp. 97–104.

121. Zhang, X. J., Wang, J., Ogorevc, B., Spichiger, U. E., 1999, "Glucose nanosensor based on Prussian-blue modified carbon-fiber cone nanoelectrode and an integrated reference electrode," *Electroanalysis*, **11**, pp. 945–949.

122. O'Halloran, M. P., Pravda, M., Guilbault, G. G., 2001, "Prussian blue bulk modified screen-printed electrodes for H_2O_2 detection and for biosensors," *Talanta*, **55**, pp. 605–611.

123. Wang, J., Rivas, G., Chicharro, M., 1996, "Iridium-dispersed carbon paste enzyme electrodes," *Electroanalysis*, **8**, pp. 434–437.

124. Dodevska, T., Horozova, E., Dimcheva, N., 2006, "Electrocatalytic reduction of hydrogen peroxide on modified graphite electrodes: application to the development of glucose biosensors," *Analytical Bioanalytical Chemistry*, **386**, pp. 1413–1418.

125. Yang, M. H., Jiang, J. H., Yang, Y. H., Chen, X. H., Shen, G. L., Yu, R. Q., 2006, "Carbon nanotube/cobalt hexacyanoferrate nanoparticle-biopolymer system for the fabrication of biosensors," *Biosensors Bioelectronics*, **21**, pp. 1791–1797.

126. Hrapovic, S., Liu, Y. L., Male, K. B., Luong, J. H. T., 2004, "Electrochemical biosensing platforms using platinum nanoparticles and carbon nanotubes," *Analytical Chemistry*, **76**, pp. 1083–1088.

127. Wang, J., Musameh, M., Lin, Y., 2003, "Solubilization of carbon nanotubes by Nafion toward the preparation of amperometric biosensors," *Journal of American Chemical Society*, **125**, pp. 2408–2409.

128. Bharathi, S., Nogami, M., 2001, "A glucose biosensor based on electrodeposited biocomposites of gold nanoparticles and glucose oxidase enzyme," *Analyst*, **126**, pp. 1919–1922.

129. Gu, T., Hasebe, Y., 2006, "DNA-Cu(II) poly(amine) complex membrane as novel catalytic layer for highly sensitive amperometric determination of hydrogen peroxide," *Biosensors Bioelectronics*, **21**, pp. 2121–2128.

130. Zhu, Z., Song, W., Burugapalli, K., Moussy, F., Li, Y. L., Zhong, X. H., 2010, "Nano-yarn carbon nanotube fiber based enzymatic glucose biosensor," *Nanotechnology*, **21**, pp. 165501.

131. Abdel-Latif, M. S., Guilbault, G. G., 1988, "Fiber-optic sensor for the determination of glucose using micellar enhanced chemiluminescence of the peroxyoxalate reaction," *Analytical Chemistry*, **60**, pp. 2671–2674.

132. Seo, H.-I., Kim, C.-S., Sohn, B.-K., Yeow, T., Son, M.-T., Haskard, M., 1997, "ISFET glucose sensor based on a new principle using the electrolysis of hydrogen peroxide," *Sensor and Actuators B – Chemistry*, **40**, pp. 1–5.

133. Park, K.-Y., Choi, S.-B., Lee, M., Sohn, B.-K., Choi, S.-Y., 2002, "ISFET glucose sensor system with fast recovery characteristics by employing electrolysis," *Sensor and Actuators B – Chemistry*, **83**, pp. 90–97.

134. Lee, C.-H., Seo, H.-I., Lee, Y.-C., Cho, B.-W., Jeong, H., Sohn, B.-K., 2000, "All solid type ISFET glucose sensor with fast response and high sensitivity characteristics," *Sensor and Actuators B – Chemistry*, **64**, pp. 37–41.

135. Besteman, K., Lee, J. O., Wiertz, F. G. M., Heering, H. A., Dekker, C., 2003, "Enzyme-coated carbon nanotubes as single-molecule biosensors," *Nano Letters*, **3**, pp. 727–730.

136. Forzani, E. S., Zhang, H. Q., Nagahara, L. A., Amlani, I., Tsui, R., Tao, N. J., 2004, "A conducting polymer nanojunction sensor for glucose detection," *Nano Letters*, **4**, pp. 1785–1788.

137. Lee, D., Cui, T., 2010, "Low-cost, transparent, and flexible single-walled carbon nanotube nanocomposite based ion-sensitive field-effect transistors for pH/glucose sensing," *Biosensors Bioelectronics*, **25**, pp. 2259–2264.

138. Bahshi, L., Freeman, R., Gill, R., Willner, I., 2009, "Optical detection of glucose by means of metal nanoparticles or semiconductor quantum dots," *Small*, **5**, pp. 676–680.

139. Gill, R., Bahshi, L., Freeman, R., Willner, I., 2008, "Optical detection of glucose and acetylcholine esterase inhibitors by H_2O_2-sensitive CdSe/ZnS quantum dots," *Angewandte Chemie International Edition*, **47**, pp. 1676–1679.

140. Hu, M., Tian, J., Lu, H. T., Weng, L. X., Wang, L. H., 2010, "H_2O_2-sensitive quantum dots for the label-free detection of glucose," *Talanta*, **82**, pp. 997–1002.

141. Li, X., Zhou, Y., Zheng, Z., Yue, X., Dai, Z., Liu, S., Tang, Z., 2009, "Glucose biosensor based on nanocomposite films of CdTe quantum dots and glucose oxidase," *Langmuir*, **25**, pp. 6580–6586.

142. Duong, H. D., Rhee, J. I., 2007, "Use of CdSe/ZnS core-shell quantum dots as energy transfer donors in sensing glucose," *Talanta*, **73**, pp. 899–905.

143. Tanne, J., Schafer, D., Khalid, W., Parak, W. J., Lisdat, F., 2011, "Light-controlled bioelectrochemical sensor based on CdSe/ZnS quantum dots," *Analytical Chemistry*, **83**, pp. 7778–7785.

144. Wu, P., He, Y., Wang, H. F., Yan, X. P., 2010, "Conjugation of glucose oxidase onto Mn-doped ZnS quantum dots for phosphorescent sensing of glucose in biological fluids," *Analytical Chemistry*, **82**, pp. 1427–1433.

145. Kim, J. H., Lim, S. Y., Nam, D. H., Ryu, J., Ku, S. H., Park, C. B., 2011, "Self-assembled, photoluminescent peptide hydrogel as a versatile platform for enzyme-based optical biosensors," *Biosensors Bioelectronics*, **26**, pp. 1860–1865.

146. Liu, Q., Lu, X., Li, J., Yao, X., Li, J., 2007, "Direct electrochemistry of glucose oxidase and electrochemical biosensing of glucose on quantum dots/carbon nanotubes electrodes," *Biosensors Bioelectronics*, **22**, pp. 3203–3209.

147. Cass, A. E., Davis, G., Francis, G. D., Hill, H. A., Aston, W. J., Higgins, I. J., Plotkin, E. V., Scott, L. D., Turner, A. P., 1984, "Ferrocene-mediated enzyme electrode for amperometric determination of glucose," *Analytical Chemistry*, **56**, pp. 667–671.

148. Guiseppi-Elie, A., Lei, C. H., Baughman, R. H., 2002, "Direct electron transfer of glucose oxidase on carbon nanotubes," *Nanotechnology*, **13**, pp. 559–564.

149. Zhao, H.-Z., Sun, J.-J., Song, J., Yang, Q.-Z., 2010, "Direct electron transfer and conformational change of glucose oxidase on carbon nanotube-based electrodes," *Carbon*, **48**, pp. 1508–1514.

150. Ivnitski, D., Artyushkova, K., Rincon, R. A., Atanassov, P., Luckarift, H. R., Johnson, G. R., 2008, "Entrapment of enzymes and carbon nanotubes in biologically synthesized silica glucose oxidase-catalyzed direct electron transfer," *Small*, **4**, pp. 357–364.

151. Wu, X. E., Zhao, F., Varcoe, J. R., Thumser, A. E., Avignone-Rossa, C., Slade, R. C. T., 2009, "Direct electron transfer of glucose oxidase immobilized in an ionic liquid reconstituted cellulose-carbon nanotube matrix," *Bioelectrochemistry*, **77**, pp. 64–68.

152. Liu, X. Q., Shi, L. H., Niu, W. X., Li, H. J., Xu, G. B., 2008, "Amperometric glucose biosensor based on single-walled carbon nanohorns," *Biosensors Bioelectronics*, **23**, pp. 1887–1890.

153. Zou, Y. J., Xiang, C. L., Sun, L. X., Xu, F., 2008, "Glucose biosensor based on electrodeposition of platinum nanoparticles onto carbon nanotubes and immobilizing enzyme with chitosan-SiO2 sol-gel," *Biosensors Bioelectronics*, **23**, pp. 1010–1016.

154. Shan, D., Zhang, J., Xue, H.-G., Ding, S.-N., Cosnier, S., 2010, "Colloidal laponite nanoparticles extended application in direct electrochemistry of glucose oxidase and reagentless glucose biosensing," *Biosensors Bioelectronics*, **25**, pp. 1427–1433.

155. Toi, V. V., Toan, N. B., Dang Khoa, T. Q., Lien Phuong, T. H., Lee, S., Choi, B., Hong, W., 2013, "Simple fabrication of exfoliated graphene/nafion hybrid as glucose bio-sensor electrodes," *4th International Conference on Biomedical Engineering*, Vietnam **40**, pp. 54–56.

156. German, N., Ramanaviciene, A., Voronovic, J., Ramanavicius, A., 2010, "Glucose biosensor based on graphite electrodes modified with glucose oxidase and colloidal gold nanoparticles," *Microchimica Acta*, **168**, pp. 221–229.

157. Švancara, I., Vytřas, K., Kalcher, K., Walcarius, A., Wang, J., 2009, "Carbon paste electrodes in facts, numbers, and notes a review on the occasion of the 50-years jubilee of carbon paste in electrochemistry and electroanalysis," *Electroanalysis*, **21**, pp. 7–28.

158. Gunasingham, H., Tan, C. H., Seow, J. K. L., 1990, "Fiber-optic glucose sensor with electrochemical generation of indicator reagent," *Analytical Chemistry*, **62**, pp. 755–759.

159. Hill, H. A. O., 1984, "Assay techniques utilising specific binding agents," European Patent No. 84303090.

160. Willner, I., Heleg-Shabtai, V., Blonder, R., Katz, E., Tao, G., Buackmann, A. F., Heller, A., 1996, "Electrical wiring of glucose oxidase by reconstitution of FAD-modified monolayers assembled onto Au-electrodes," *Journal of American Chemical Society*, **118**, pp. 10321–10322.

161. Patolsky, F., Weizmann, Y., Willner, I., 2004, "Long-range electrical contacting of redox enzymes by SWCNT connectors," *Angewandte Chemie International Edition*, **43**, pp. 2113–2117.

162. Liu, J., Chou, A., Rahmat, W., Paddon-Row, M., Gooding, J., 2005, "Achieving direct electrical connection to glucose oxidase using aligned single walled carbon nanotube arrays," *Electroanalysis*, **17**, pp. 38–46.

163. Zheng, J., He, Y., Sheng, Q., Zhang, H., 2011, "DNA as a linker for biocatalytic deposition of Au nanoparticles on graphene and its application in glucose detection," *Journal of Materials Chemistry*, **21**, pp. 12873–12879.

164. Chen, Y., Li, Y., Sun, D., Tian, D., Zhang, J., Zhu, J.-J., 2011, "Fabrication of gold nanoparticles on bilayer graphene for glucose electrochemical biosensing," *Journal of Materials Chemistry*, **21**, pp. 7604–7611.

165. Zhou, K., Zhu, Y., Yang, X., Li, C., 2010, "Electrocatalytic oxidation of glucose by the glucose oxidase immobilized in graphene-Au-Nafion biocomposite," *Electroanalysisysis*, **22**, pp. 259–264.

166. Adiga, S. P., Jin, C., Curtiss, L. A., Monteiro-Riviere, N. A., Narayan, R. J., 2009, "Nanoporous membranes for medical and biological applications," *WIREs Nanomedicine Nanobiotechnology*, B, pp. 568–581.

167. Ulbricht, M., 2006, "Advanced functional polymer membranes," *Polymer*, **47**, pp. 2217–2262.

168. Langley, P. J., Hulliger, J., 1999, "Nanoporous and mesoporous organic structures new openings for materials research," *Chemical Society Reviews*, **28**, pp. 279–291.

169. Updike, S. J., Hicks, G. P., 1967, "The enzyme electrode," *Nature*, **214**, pp. 986–988.

170. Yu, B., Moussy, Y., Moussy, F., 2005, "Coil-type implantable glucose biosensor with excess enzyme loading," *Frontiers in Biosciences*, **10**, pp. 512–520.

171. Li, Q., Luo, G., Feng, J., Zhou, Q., Zhang, L., Zhu, Y., 2001, "Amperometric detection of glucose with glucose oxidase absorbed on porous nanocrystalline TiO_2 film," *Electroanalysis*, **13**, pp. 413–416.

172. Singh, S. P., Arya, S. K., Pandey, P., Malhotra, B. D., Saha, S., Sreenivas, K., Gupta, V., 2007, "Cholesterol biosensor based on rf sputtered zinc oxide nanoporous thin film," *Applied Physics Letters*, **91**, pp. 063901.

173. Saha, S., Arya, S. K., Singh, S. P., Sreenivas, K., Malhotra, B. D., Gupta, V., 2009, "Nanoporous cerium oxide thin film for glucose biosensor," *Biosensors Bioelectronics*, **24**, pp. 2040–2045.

174. Saha, S., Arya, S. K., Singh, S. P., Sreenivas, K., Malhotra, B. D., Gupta, V., 2009, "Zinc oxide-potassium ferricyanide composite thin film matrix for biosensing applications," *Analytica Chimica Acta*, **653**, pp. 212–216.

175. Liu, H., Hu, N., 2007, "Study on direct electrochemistry of glucose oxidase stabilized by cross-linking and immobilized in silica nanoparticle films," *Electroanalysis*, **19**, pp. 884–892.

176. Malitesta, C., Palmisano, F., Torsi, L., Zambonin, P. G., 1990, "Glucose fast-response amperometric sensor based on glucose-oxidase immobilized in an electropolymerized poly(ortho-phenylenediamine) film," *Analytical Chemistry*, **62**, pp. 2735–2740.

177. Netchiporouk, L. I., Shram, N. F., Jaffrezic-Renault, N., Martelet, C., Cespuglio, R., 1996, "In vivo brain glucose measurements differential normal pulse voltammetry with enzyme-modified carbon fiber microelectrodes," *Analytical Chemistry*, **68**, pp. 4358–4364.

178. Cosnier, S., 1999, "Biomolecule immobilization on electrode surfaces by entrapment or attachment to electrochemically polymerized films. A review," *Biosensors Bioelectronics*, **14**, pp. 443–456.

179. Pandey, P. C., Upadhyay, S., Pathak, H. C., 1999, "A new glucose sensor based on encapsulated glucose oxidase within organically modified sol-gel glass," *Sensor and Actuators B - Chemistry*, **60**, pp. 83–89.

180. Tatsu, Y., Yamashita, K., Yamaguchi, M., Yamamura, S., Yamamoto, H., Yoshikawa, S., 1992, "Entrapment of glucose oxidase in silica gel by the sol-gel method and its application to glucose sensor," *Chemistry Letters*, **21**, pp. 1615–1618.

181. Ansari, A. A., Solanki, P. R., Malhotra, B. D., 2008, "Sol-gel derived nanostructured cerium oxide film for glucose sensor," *Applied Physics Letters*, **92**, pp. 263901–263903.

182. Yu, J., Liu, S., Ju, H., 2003, "Glucose sensor for flow injection analysis of serum glucose based on immobilization of glucose oxidase in titania sol-gel membrane," *Biosensors Bioelectronics*, **19**, pp. 401–409.

183. Zhao, W., Xu, J. J., Chen, H. Y., 2006, "Electrochemical biosensors based on layer-by-layer assemblies," *Electroanalysis*, **18**, pp. 1737–1748.

184. Hou, S. F., Yang, K. S., Fang, H. Q., Chen, H. Y., 1998, "Amperometric glucose enzyme electrode by immobilizing glucose oxidase in multilayers on self-assembled monolayers surface," *Talanta*, **47**, pp. 561–567.

185. Hou, S.-F., Fang, H.-Q., Chen, H.-Y., 1997, "An amperometric enzyme electrode for glucose using immobilized glucose oxidase in a ferrocene attached poly(4-vinylpyridine) multilayer film," *Analytical Letters*, **30**, pp. 1631–1641.

186. Calvo, E. J., Danilowicz, C., Wolosiuk, A., 2002, "Molecular "wiring" enzymes in organized nanostructures," *Journal of American Chemical Society*, **124**, pp. 2452–2453.

187. Hodak, J., Etchenique, R., Calvo, E. J., Singhal, K., Bartlett, P. N., 1997, "Layer-by-layer self-assembly of glucose oxidase with a poly(allylamine) ferrocene redox mediator," *Langmuir*, **13**, pp. 2708–2716.

188. Liu, J., Tian, S., Knoll, W., 2005, "Properties of polyaniline/carbon nanotube multilayer films in neutral solution and their application for stable low-potential detection of reduced Î²-nicotinamide adenine dinucleotide," *Langmuir*, **21**, pp. 5596–5599.

189. Hoshi, T., Anzai, J.-I., Osa, T., 1995, "Controlled deposition of glucose oxidase on platinum electrode based on an avidin/biotin system for the regulation of output current of glucose sensors," *Analytical Chemistry*, **67**, pp. 770–774.

190. Bourdillon, C., Demaille, C., Moiroux, J., Saveant, J.-M., 1994, "Step-by-step immunological construction of a fully active multilayer enzyme electrode," *Journal of American Chemical Society*, **116**, pp. 10328–10329.

191. Sirkar, K., Revzin, A., Pishko, M. V., 2000, "Glucose and lactate biosensors based on redox polymer/oxidoreductase nanocomposite thin films," *Analytical Chemistry*, **72**, pp. 2930–2936.

192. Wang, Z. G., Wano, Y., Xu, H., Li, G., Xu, Z. K., 2009, "Carbon nanotube-filled nanofibrous membranes electrospun from poly(acrylonitrile-co-acrylic acid) for glucose biosensor," *Journal of Physical Chemistry – C*, **113**, pp. 2955–2960.

193. Ren, G., Xu, X., Liu, Q., Cheng, J., Yuan, X., Wu, L., Wan, Y., 2006, "Electrospun poly(vinyl alcohol)/glucose oxidase biocomposite membranes for biosensor applications," *Reactive Functional Polymers*, **66**, pp. 1559–1564.

194. Doretti, L., Ferrara, D., Gattolin, P., Lora, S., Schiavon, F., Veronese, F. M., 1998, "PEG-modified glucose oxidase immobilized on a PVA cryogel membrane for amperometric biosensor applications," *Talanta*, **45**, pp. 891–898.

195. Manesh, K. M., Kim, H. T., Santhosh, P., Gopalan, A. I., Lee, K. P., 2008, "A novel glucose biosensor based on immobilization of glucose oxidase into multiwall carbon nanotubes-polyelectrolyte-loaded electrospun nanofibrous membrane," *Biosensors Bioelectronics*, **23**, pp. 771–779.

196. Liu, Y., Chen, J., Anh, N. T., Too, C. O., Misoska, V., Wallace, G. G., 2008, "Nanofiber mats from DNA, SWNTs, and poly(ethylene oxide) and their application in glucose biosensors," *Journal of Electrochemical Society*, **155**, pp. K100–K103.

197. Aussawasathien, D., Dong, J. H., Dai, L., 2005, "Electrospun polymer nanofiber sensors," *Synthetic Metals*, **154**, pp. 37–40.

198. Jia, W.-Z., Wang, K., Xia, X.-H., 2010, "Elimination of electrochemical interferences in glucose biosensors," *TrAC Trends Analytical Chemistry*, **29**, pp. 306–318.

199. Choi, S. H., Lee, S. D., Shin, J. H., Ha, J., Nam, H., Cha, G. S., 2002, "Amperometric biosensors employing an insoluble oxidant as an interference-removing agent," *Analytica Chimica Acta*, **461**, pp. 251–260.

200. Cui, G., Kim, S. J., Choi, S. H., Nam, H., Cha, G. S., Paeng, K.-J., 2000, "A disposable amperometric sensor screen printed on a nitrocellulose strip a glucose biosensor employing lead oxide as an interference-removing agent," *Analytical Chemistry*, **72**, pp. 1925–1929.

201. Xu, J.-J., Luo, X.-L., Du, Y., Chen, H.-Y., 2004, "Application of MnO_2 nanoparticles as an eliminator of ascorbate interference to amperometric glucose biosensors," *Electrochemical Communications*, **6**, pp. 1169–1173.

202. Sasso, S. V., Pierce, R. J., Walla, R., Yacynych, A. M., 1990, "Electro-polymerized 1,2-diaminobenzene as a means to prevent interferences and fouling and to stabilize immobilized enzyme in electrochemical biosensors," *Analytical Chemistry*, **62**, pp. 1111–1117.

203. Palmisano, F., Centonze, D., Guerrieri, A., Zambonin, P. G., 1993, "An interference-free biosensor based on glucose-oxidase electrochemically immobilized in a nonconducting poly(pyrrole) film for continuous subcutaneous monitoring of glucose through microdialysis sampling," *Biosensors Bioelectronics*, **8**, pp. 393–399.

204. Lukachova, L. V., Kotenikova, E. A., D'Ottavi, D., Shkerin, E. A., Karyakina, E. E., Moscone, D., Palleschi, G., Curulli, A., Karyakin, A. A., 2002, "Electrosynthesis of poly-o-diaminobenzene on the Prussian blue modified electrodes for improvement of hydrogen peroxide transducer characteristics," *Bioelectrochemistry*, **55**, pp. 145–148.

205. Zhang, Z., Liu, H., Deng, J., 1996, "A glucose biosensor based on immobilization of glucose oxidase in electropolymerized o-aminophenol film on platinized glassy carbon electrode," *Analytical Chemistry*, **68**, pp. 1632–1638.

206. Chen, X., Matsumoto, N., Hu, Y., Wilson, G. S., 2002, "Electrochemically mediated electrodeposition/electropolymerization to yield a glucose microbiosensor with improved characteristics," *Analytical Chemistry*, **74**, pp. 368–372.

207. Murphy, L. J., 1998, "Reduction of interference response at a hydrogen peroxide detecting electrode using electropolymerized films of substituted naphthalenes," *Analytical Chemistry*, **70**, pp. 2928–2935.

208. Emr, S. A., Yacynych, A. M., 1995, "Use of polymer-films in amperometric biosensors," *Electroanalysis*, **7**, pp. 913–923.

209. Sternberg, R., Bindra, D. S., Wilson, G. S., Thevenot, D. R., 1988, "Covalent enzyme coupling on cellulose-acetate membranes for glucose sensor development," *Analytical Chemistry*, **60**, pp. 2781–2786.

210. Muguruma, H., Hiratsuka, A., Karube, I., 2000, "Thin-film glucose biosensor based on plasma-polymerized film a simple design for mass production," *Analytical Chemistry*, **72**, pp. 2671–2675.

211. Jung, S.-K., Wilson, G. S., 1996, "Polymeric mercaptosilane-modified platinum electrodes for elimination of interferants in glucose biosensors," *Analytical Chemistry*, **68**, pp. 591–596.

212. Moussy, F., Jakeway, S., Harrison, D. J., Rajotte, R. V., 1994, "In vitro and in vivo performance and lifetime of perfluorinated ionomer-coated glucose sensors after high-temperature curing," *Analytical Chemistry*, **66**, pp. 3882–3888.

213. Wang, J., Wu, H., 1993, "Permselective lipid poly(o-phenylenediamine) coatings for amperometric biosensing of glucose," *Analytica Chimica Acta*, **283**, pp. 683–688.

214. Zhang, Y. N., Hu, Y. B., Wilson, G. S., Moattisirat, D., Poitout, V., Reach, G., 1994, "Elimination of the acetaminophen interference in an implantable glucose sensor," *Analytical Chemistry*, **66**, pp. 1183–1188.

215. Vaidya, R., Atanasov, P., Wilkins, E., 1995, "Effect of interference on the performance of glucose enzyme electrodes using Nafion® coatings," *Medical Engineering Physics*, **17**, pp. 416–424.

216. Rodríguez, M., Rivas, G., 2004, "Assembly of glucose oxidase and different polyelectrolytes by means of electrostatic layer-by-layer adsorption on thiolated gold surface," *Electroanalysis*, **16**, pp. 1717–1722.

217. Shults, M. C., Capelli, C. C., Updike, S. J., 1991, "Biological fluid measuring device," US Patent No. 4994167.
218. Updike, S. J., Shults, M., Ekman, B., 1982, "Implanting the glucose enzyme electrode: problems, progress, and alternative solutions," *Diabetes Care*, **5**, pp. 207–212.
219. Wang, N., Burugapalli, K., Song, W., Halls, J., Moussy, F., Zheng, Y., Ma, Y., Wu, Z., Li, K., 2013, "Tailored fibro-porous structure of electrospun polyurethane membranes, their size-dependent properties and trans-membrane glucose diffusion," *Journal of Membrane Science*, **427**, pp. 207–217.
220. Wang, N., Burugapalli, K., Song, W., Halls, J., Moussy, F., Ray, A., Zheng, Y., 2013, "Electrospun fibro-porous polyurethane coatings for implantable glucose biosensors," *Biomaterials*, **34**, pp. 888–901.
221. Matsumoto, T., Furusawa, M., Fujiwara, H., Matsumoto, Y., Ito, N., 1998, "A micro-planar amperometric glucose sensor unsusceptible to interference species," *Sensor and Actuators B – Chemistry*, **49**, pp. 68–72.
222. Ward, W. K., Jansen, L. B., Anderson, E., Reach, G., Klein, J.-C., Wilson, G. S., 2002, "A new amperometric glucose microsensor: in vitro and short-term in vivo evaluation," *Biosensors Bioelectronics*, **17**, pp. 181–189.
223. Linke, B., Kerner, W., Kiwit, M., Pishko, M., Heller, A., 1994, "Amperometric biosensor for in vivo glucose sensing based on glucose oxidase immobilized in a redox hydrogel," *Biosensors Bioelectronics*, **9**, pp. 151–158.
224. Matsumoto, T., Ohashi, A., Ito, N., Fujiwara, H., Matsumoto, T., 2001, "A long-term lifetime amperometric glucose sensor with a perfluorocarbon polymer coating," *Biosensors Bioelectronics*, **16**, pp. 271–276.
225. Reddy, S. M., Vadgama, P., 2002, "Entrapment of glucose oxidase in non-porous poly(vinyl chloride)," *Analytica Chimica Acta*, **461**, pp. 57–64.
226. Chung, T. D., 2002, "In vitro evaluation of the continuous monitoring glucose sensors with perfluorinated tetrafluoroethylene coatings," *Bulletin of Korean Chemical Society*, **24**, pp. 514–516.
227. Moussy, F., Harrison, D. J., Rajotte, R. V., 1994, "A miniaturized Nafion-based glucose sensor in vitro and in vivo evaluation in dogs," *International Journal of Artificial Organs*, **17**, pp. 88–94.
228. Mercado, R. C., Moussy, F., 1998, "In vitro and in vivo mineralization of Nafion membrane used for implantable glucose sensors," *Biosensors Bioelectronics*, **13**, pp. 133–145.
229. Brauker, J. H., Carr-Brendel, V., Tapsak, M. A., 2012, "Porous membranes for use with implantable devices," US Patent No. 8118877.
230. Madaras, M. B., Popescu, I. C., Ufer, S., Buck, R. P., 1996, "Microfabricated amperometric creatine and creatinine biosensors," *Analytica Chimica Acta*, **319**, pp. 335–345.
231. Trzebinski, J., Moniz, A. R.-B., Sharma, S., Burugapalli, K., Moussy, F., Cass, A. E. G., 2011, "Hydrogel membrane improves batch-to-batch reproducibility of an enzymatic glucose biosensor," *Electroanalysis*, **23**, pp. 2789–2795.
232. Yu, B., Long, N., Moussy, Y., Moussy, F., 2006, "A long-term flexible minimally-invasive implantable glucose biosensor based on an epoxy-enhanced polyurethane membrane," *Biosensors Bioelectronics*, **21**, pp. 2275–2282.

233. Yu, B., Moussy, Y., Moussy, F., 2005, "Lifetime improvement of glucose biosensor by epoxy-enhanced PVC membrane," *Electroanalysis*, **17**, pp. 1771–1779.
234. Uehara, H., Kakiage, M., Sekiya, M., Sakuma, D., Yamonobe, T., Takano, N., Barraud, A., Meurville, E., Ryser, P., 2009, "Size-selective diffusion in nanoporous but flexible membranes for glucose sensors," *ACS Nano*, **3**, pp. 924–932.
235. Chu, M., Kudo, H., Shirai, T., Miyajima, K., Saito, H., Morimoto, N., Yano, K., Iwasaki, Y., Akiyoshi, K., Mitsubayashi, K., 2009, "A soft and flexible biosensor using a phospholipid polymer for continuous glucose monitoring," *Biomedical Microdevices*, **11**, pp. 837–842.
236. Chen, C.-Y., Ishihara, K., Nakabayashi, N., Tamiya, E., Karube, I., 1999, "Multifunctional biocompatible membrane and its application to fabricate a miniaturized glucose sensor with potential for use in vivo," *Biomedical Microdevices*, **1**, pp. 155–166.
237. Moussy, F., Jakeway, S., Harrison, D. J., Rajotte, R. V., 1994, "In-vitro and in-vivo performance and lifetime of perfluorinated ionomer-coated glucose sensors after high-temperature curing," *Analytical Chemistry*, **66**, pp. 3882–3888.
238. Harrison, D. J., Turner, R. F., Baltes, H. P., 1988, "Characterization of perfluorosulfonic acid polymer coated enzyme electrodes and a miniaturized integrated potentiostat for glucose analysis in whole blood," *Analytical Chemistry*, **60**, pp. 2002–2007.
239. Bruck, S. D., 1974, *Blood Compatible Synthetic Polymers*. Charles, C. T., ed, Springfield, IL, pp. 94.
240. Valdes, T. I., Moussy, F., 1999, "A ferric chloride pre-treatment to prevent calcification of Nafion membrane used for implantable biosensors," *Biosensors Bioelectronics*, **14**, pp. 579–585.
241. Galeska, I., Chattopadhyay, D., Moussy, F., Papadimitrakopoulos, F., 2000, "Calcification-resistant Nafion/Fe3+ assemblies for implantable biosensors," *Biomacromolecules*, **1**, pp. 202–207.
242. Yu, B., Ju, Y., West, L., Moussy, Y., Moussy, F., 2007, "An investigation of long-term performance of minimally invasive glucose biosensors," *Diabetes Technology and Therapeutics*, **9**, pp. 265–275.
243. Mang, A., Pill, J., Gretz, N., Kranzlin, B., Buck, H., Schoemaker, M., Petrich, W., 2005, "Biocompatibility of an electrochemical sensor for continuous glucose monitoring in subcutaneous tissue," *Diabetes Technology and Therapeutics*, **7**, pp. 163–173.
244. Galeska, I., Hickey, T., Moussy, F., Kreutzer, D., Papadimitrakopoulos, F., 2001, "Characterization and biocompatibility studies of novel humic acids based films as membrane material for an implantable glucose sensor," *Biomacromolecules*, **2**, pp. 1249–1255.
245. Tipnis, R., Vaddiraju, S., Jain, F., Burgess, D. J., Papadimitrakopoulos, F., 2007, "Layer-by-layer assembled semipermeable membrane for amperometric glucose sensors," *Journal of Diabetes Science and Technology*, **1**, pp. 193–200.

246. Wisniewski, N., Klitzman, B., Miller, B., Reichert, W. M., 2001, "Decreased analyte transport through implanted membranes differentiation of biofouling from tissue effects," *Journal of Biomedical Materials Research*, **57**, pp. 513–521.

247. Brauker, J. H., Shults, M. C., Tapsak, M. A., 2004, "Membrane for use with implantable devices," US Patent No. 6702857.

248. Wisniewski, N., Moussy, F., Reichert, W. M., 2000, "Characterization of implantable biosensor membrane biofouling," *Fresenius J Analytical Chemistry*, **366**, pp. 611–621.

249. Wisniewski, N., Reichert, M., 2000, "Methods for reducing biosensor membrane biofouling," *Colloids and Surfaces – B*, **18**, pp. 197–219.

250. Vaddiraju, S., Tomazos, I., Burgess, D. J., Jain, F. C., Papadimitrakopoulos, F., 2010, "Emerging synergy between nanotechnology and implantable biosensors a review," *Biosensors Bioelectronics*, **25**, pp. 1553–1565.

251. Tang, L., Wu, Y., Timmons, R. B., 1998, "Fibrinogen adsorption and host tissue responses to plasma functionalized surfaces," *Journal of Biomedical Materials Research*, **42**, pp. 156–163.

252. Barbosa, J. N., Madureira, P., Barbosa, M. A., Águas, A. P., 2006, "The influence of functional groups of self-assembled monolayers on fibrous capsule formation and cell recruitment," *Journal of Biomedical Materials Research*, **76A**, pp. 737–743.

253. Krishnan, S., Weinman, C. J., Ober, C. K., 2008, "Advances in polymers for anti-biofouling surfaces," *Journal of Materials Chemistry*, **18**, pp. 3405–3413.

254. Moussy, F., Harrison, D. J., O'Brien, D. W., Rajotte, R. V., 1993, "Performance of subcutaneously implanted needle-type glucose sensors employing a novel trilayer coating," *Analytical Chemistry*, **65**, pp. 2072–2077.

255. Yu, B., Wang, C., Ju, Y. M., West, L., Harmon, J., Moussy, Y., Moussy, F., 2008, "Use of hydrogel coating to improve the performance of implanted glucose sensors," *Biosensors Bioelectronics*, **23**, pp. 1278–1284.

256. Dungel, P., Long, N., Yu, B., Moussy, Y., Moussy, F., 2008, "Study of the effects of tissue reactions on the function of implanted glucose sensors," *Journal of Biomedical Materials Research*, **85A**, pp. 699–706.

257. Quinn, C. A., Connor, R. E., Heller, A., 1997, "Biocompatible, glucose-permeable hydrogel for in situ coating of implantable biosensors," *Biomaterials*, **18**, pp. 1665–1670.

258. McKinley, B. A., Wong, K. C., Janata, J., Jordan, W. S., Westenskow, D. R., 1981, "In vivo continuous monitoring of ionized calcium in dogs using ion sensitive field effect transistors," *Critical Care Medicine*, **9**, pp. 333–339.

259. Margules, G. S., Hunter, C. M., MacGregor, D. C., 1983, "Hydrogel based in vivo reference electrode catheter," *Medical & Biological Engineering & Computing*, **21**, pp. 1–8.

260. Shimada, K., Yano, M., Shibatani, K., Komoto, Y., Esashi, M., Matsuo, T., 1980, "Application of catheter-tip i.s.f.e.t. for continuous in vivo measurement," *Medical & Biological Engineering & Computing*, **18**, pp. 741–745.

261. Ishihara, K., Tanaka, S., Furukawa, N., Nakabayashi, N., Kurita, K., 1996, "Improved blood compatibility of segmented polyurethanes by polymeric additives having phospholipid polar groups. I. Molecular design of polymeric

additives and their functions," *Journal of Biomedical Materials Research*, **32**, pp. 391–399.

262. Yang, Y., Zhang, S. F., Kingston, M. A., Jones, G., Wright, G., Spencer, S. A., 2000, "Glucose sensor with improved haemocompatibilty," *Biosensors Bioelectronics*, **15**, pp. 221–227.

263. Ishihara, K., Nakabayashi, N., Nishida, K., Sakakida, M., Shichiri, M., 1994, "New biocompatible polymer. diagnostic biosensor polymers," *ACS Symposium Series*, **556**, pp. 194–210.

264. Praveen, S. S., Hanumantha, R., Belovich, J. M., Davis, B. L., 2003, "Novel hyaluronic acid coating for potential use in glucose sensor design," *Diabetes Technology and Therapeutics*, **5**, pp. 393–399.

265. Saha, K., Pollock, J. F., Schaffer, D. V., Healy, K. E., 2007, "Designing synthetic materials to control stem cell phenotype," *Current Opinion in Chemical Biology*, **11**, pp. 381–387.

266. Lopez, C. A., Fleischman, A. J., Roy, S., Desai, T. A., 2006, "Evaluation of silicon nanoporous membranes and ECM-based microenvironments on neurosecretory cells," *Biomaterials*, **27**, pp. 3075–3083.

267. Popat, K. C., Eltgroth, M., LaTempa, T. J., Grimes, C. A., Desai, T. A., 2007, "Decreased staphylococcus epidermis adhesion and increased osteoblast functionality on antibiotic-loaded titania nanotubes," *Biomaterials*, **28**, pp. 4880–4888.

268. Sahlin, H., Contreras, R., Gaskill, D. F., Bjursten, L. M., Frangos, J. A., 2006, "Anti-inflammatory properties of micropatterned titanium coatings," *Journal of Biomedical Materials Research*, **77A**, pp. 43–49.

269. Ainslie, K. M., Tao, S. L., Popat, K. C., Desai, T. A., "In vitro immunogenicity of silicon-based micro- and nanostructured surfaces," *ACS Nano*, 2008, **2**, pp. 1076.

270. Desai, T. A., Hansford, D. J., Leoni, L., Essenpreis, M., Ferrari, M., 2000, "Nanoporous anti-fouling silicon membranes for biosensor applications," *Biosensors Bioelectronics*, **15**, pp. 453–462.

271. Striemer, C. C., Gaborski, T. R., McGrath, J. L., Fauchet, P. M., 2007, "Charge- and size-based separation of macromolecules using ultrathin silicon membranes" *Nature*, **445**, pp. 749–753.

272. Narayan, R. J., Aggarwal, R., Wei, W., Jin, C., Monteiro-Riviere, N. A., Crombez, R., Shen, W., 2008, "Mechanical and biological properties of nanoporous carbon membranes," *Biomedical Materials*, **3**, pp. 034107.

273. Khun, N. W., Liu, E., Yang, G. C., Ma, W. G., Jiang, S. P., 2009, "Structure and corrosion behavior of platinum/ruthenium/nitrogen doped diamond-like carbon thin films," *Journal of Applied Physics*, **106**, pp. 13506–13515.

274. Maalouf, R., Chebib, H., Saikali, Y., Vittori, O., Sigaud, M., Garrelie, F., Donnet, C., Jaffrezic-Renault, N., 2007, "Characterization of different diamond-like carbon electrodes for biosensor design," *Talanta*, **72**, pp. 310–314.

275. Onuki, Y., Bhardwaj, U., Papadimitrakopoulos, F., Burgess, D. J., 2008, "A review of the biocompatibility of implantable devices current challenges to overcome foreign body response," *Journal of Diabetes Science and Technology*, **2**, pp. 1003–1015.

276. Brauker, J. H., Carr-Brendel, V. E., Martinson, L. A., Crudele, J., Johnston, W. D., Johnson, R. C., 1995, "Neovascularization of synthetic membranes directed by membrane microarchitecture," *Journal of Biomedical Materials Research*, **29**, pp. 1517–1524.

277. Koschwanez, H. E., Yap, F. Y., Klitzman, B., Reichert, W. M., 2008, "In vitro and in vivo characterization of porous poly-L-lactic acid coatings for subcutaneously implanted glucose sensors," *Journal of Biomedical Materials Research*, **87**, pp. 792–807.

278. Sharkawy, A. A., Klitzman, B., Truskey, G. A., Reichert, W. M., 1997, "Engineering the tissue which encapsulates subcutaneous implants. I. Diffusion properties," *Journal of Biomedical Materials Research*, **37**, pp. 401–412.

279. Sharkawy, A. A., Klitzman, B., Truskey, G. A., Reichert, W. M., 1998, "Engineering the tissue which encapsulates subcutaneous implants. III. Effective tissue response times," *Journal of Biomedical Materials Research*, **40**, pp. 598–605.

280. Sharkawy, A. A., Klitzman, B., Truskey, G. A., Reichert, W. M., 1998, "Engineering the tissue which encapsulates subcutaneous implants. II. Plasma-tissue exchange properties," *Journal of Biomedical Materials Research*, **40**, pp. 586–597.

281. Ward, W. K., Slobodzian, E. P., Tiekotter, K. L., Wood, M. D., 2002, "The effect of microgeometry, implant thickness and polyurethane chemistry on the foreign body response to subcutaneous implants," *Biomaterials*, **23**, pp. 4185–4192.

282. Shults, M. C., Updike, S. J., Rhodes, R. K., Gilligan, B. J., Tapsak, M. A., 2005, "Device and method for determining analyte levels," US Patent No. 6862465.

283. Ju, Y. M., Yu, B., Koob, T. J., Moussy, Y., Moussy, F., 2008, "A novel porous collagen scaffold around an implantable biosensor for improving biocompatibility. I. In vitro/in vivo stability of the scaffold and in vitro sensitivity of the glucose sensor with scaffold," *Journal of Biomedical Materials Research*, **87A**, pp. 136–146.

284. Ju, Y. M., Yu, B., West, L., Moussy, Y., Moussy, F., 2010, "A novel porous collagen scaffold around an implantable biosensor for improving biocompatibility. II. Long-term in vitro/in vivo sensitivity characteristics of sensors with NDGA- or GA-crosslinked collagen scaffolds," *Journal of Biomedical Materials Research*, **92A**, pp. 650–658.

285. Sanders, J. E., Cassisi, D. V., Neumann, T., Golledge, S. L., Zachariah, S. G., Ratner, B. D., Bale, S. D., 2003, "Relative influence of polymer fiber diameter and surface charge on fibrous capsule thickness and vessel density for single-fiber implants," *Journal of Biomedical Materials Research*, **65A**, pp. 462–467.

286. Burugapalli, K., Wang, N., Wijesuriya, S., Song, W., Moussy, F., 2012, "Electrospun membrane coated glucose biosensor – in vivo efficacy," *9th World Biomaterials Congress*, Chengdu, China.

287. "Diabetes Atlas", 2009, 4th edn, International Diabetes Federation, Brussels.

288. Malmberg, K., Ryden, L., Mellbin, L., 2010, "Managing cardiovascular risk in patients with type 2 diabetes – challenges and opportunities," *European Cardiology*, **6**, pp. 31–35.

289. Diabetes.org, "Diabetes-Prevalence", 2013, http//www.diabetes.org.uk/Professionals/Publications-reports-and-resources/Reports-statistics-and-case-studies/Reports/Diabetes-prevalence-2012-March-2013/.

290. Cash, K. J., Clark, H. A., 2010, "Nanosensors and nanomaterials for monitoring glucose in diabetes," *Trends in Molecular Medicine*, **16**, pp. 584–593.

291. Zenkl, G., Mayr, T., Klimant, I., 2008, "Sugar-responsive fluorescent nano-spheres," *Macromolecular Bioscience*, **8**, pp. 146–152.

292. Zenkl, G., Klimant, I., 2009, "Fluorescent acrylamide nanoparticles for bo-ronic acid based sugar sensing from probes to sensors," *Microchimica Acta*, **166**, pp. 123–131.

293. Billingsley, K., Balaconis, M. K., Dubach, J. M., Zhang, N., Lim, E., Francis, K. P., Clark, H. A., 2010. "Fluorescent nano-optodes for glucose detection," *Analytical Chemistry*, **82**, pp. 3707–3713.

294. Saxl, T., Khan, F., Matthews, D. R., Zhi, Z. L., Rolinski, O., Ameer-Beg, S., Pickup, J., 2009, "Fluorescence lifetime spectroscopy and imaging of nano-engineered glucose sensor microcapsules based on glucose/galactose-binding protein," *Biosensors Bioelectronics*, **24**, pp. 3229–3234.

295. Chinnayelka, S., McShane, M. J., 2004, "Glucose-sensitive nanoassemblies comprising affinity-binding complexes trapped in fuzzy microshells," *Journal of Fluorescence*, **14**, pp. 585–595.

296. Chinnayelka, S., Zhu, H., McShane, M., 2008, "Near-infrared resonance energy transfer glucose biosensors in hybrid microcapsule carriers," *Journal of Sensors*, **2008**, 346016 pp. 1–11.

297. Scodeller, P., Flexer, V., Szamocki, R., Calvo, E. J., Tognalli, N., Troiani, H., Fainstein, A., 2008, "Wired-enzyme core-shell Au nanoparticle biosensor," *Journal of American Chemical Society*, **130**, pp. 12690–12697.

298. Lyandres, O., Yuen, J. M., Shah, N. C., Van Duyne, R. P., Walsh, J. T., Glucksberg, M. R., 2008, "Progress toward an in vivo surface-enhanced Raman spectroscopy glucose sensor," *Diabetes Technology and Therapeutics*, **10**, pp. 257–265.

299. Barone, P. W., Strano, M. S., 2009, "Single walled carbon nanotubes as reporters for the optical detection of glucose," *Journal of Diabetes Science and Technology*, **3**, pp. 242–252.

300. Barone, P. W., Yoon, H., Ortiz-Garcia, R., Zhang, J., Ahn, J. H., Kim, J. H., Strano, M. S., 2009, "Modulation of single-walled carbon nanotube photo-luminescence by hydrogel swelling," *ACS Nano*, **3**, pp. 3869–3877.

301. Li, Y., Liu, X., Yuan, H., Xiao, D., 2009, "Glucose biosensor based on the room-temperature phosphorescence of TiO_2/SiO_2 nanocomposite," *Biosensors Bioelectronics*, **24**, pp. 3706–3710.

302. Tominaga, M., Shimazoe, T., Nagashima, M., Taniguchi, I., 2005, "Electrocatalytic oxidation of glucose at gold nanoparticle-modified carbon electrodes in alkaline and neutral solutions," *Electrochemical Communications*, **7**, pp. 189–193.

303. Holt-Hindle, P., Nigro, S., Asmussen, M., Chen, A., 2008, "Amperometric glucose sensor based on platinum-iridium nanomaterials," *Electrochemical Communications*, **10**, pp. 1438–1441.

www.ingramcontent.com/pod-product-compliance
Lightning Source LLC
Chambersburg PA
CBHW050125240326
41458CB00122B/1421